ASPHALT

by

PAL ZAKAR

CHEMICAL PUBLISHING COMPANY, INC.

New York

1971

1971

CHEMICAL PUBLISHING COMPANY, INC.

Library of Congress Catalogue Card Number: 73-122761

Original title of book
BITUMEN

Printed in the United States of America

Preface

The American edition of the present book seemed to be justified by the interest taken in it by the European public. The good reception of the book was based on the joint treatment of asphalt production problems and the ever-increasing quality requirements. Dealing with the subject in this way corresponds to the original purpose of the book. Therefore this edition is also based on the structure of the first edition which need not have been altered. However, the text had to be complemented by certain recent findings. Considerable additions had to be included in the chapters dealing with asphalt emulsion production and asphalt uses.

The quality requirements in asphalt use are shown by the standard specifications and recommendations of the individual countries. Some recent U.S. specifications have also been treated in the present edition. It must be emphasized that only the most important characteristics of the standards have been taken up to maintain the original purpose of the book. It was not desirable to give full particulars on standards, since those can easily be found in the given references if the necessity arises.

As for the crude oils and asphalts taken up in the present book, they are typical materials. For example, the Hungarian crude oil and asphalt from Nagylengyel are essentially similar to certain American crude oils with high sulphur and asphalt content. As a consequence, production correlations and references relate to all similar asphaltic crude oils of American or any other origin. The same applies, of course, to other crude oils having lower asphalt contents which also serve as examples.

The translation and completion of the American edition has been promoted to a great extent by the valuable assistance of The Asphalt Institute, especially by permitting the author to adopt certain statements and data covering asphalt uses and quality specifications. Special thanks are due for this.

P. Zakar

Contents

1 Production and Properties of Asphalt

1.1 Nomenclature and Origin

Many thousand years before crude oil exploration and its industrial processing began, man had recognised the numerous advantages of asphalt application and had started the production of bituminous materials found in natural deposits, using the available methods[1,2,3]. In the course of history, more and more asphalt deposits were detected. In the 18th century the importance of asphalt consumption increased by its extensive application in road building. In the 19th century the development of the asphalt industry reached its height. This was joined by the asphalt production based on the systematic industrial processing of crude oil in the second half of the 19th century, since natural asphalt deposits could no longer meet the asphalt requirements, which had increased to a great extent. At present the exploitation of the known deposits and of the recently found natural deposits is still in progress, however the quantities produced answer only part of the world's demand for asphalt.

Natural asphalts and their early applications, as well as the industrial asphalt production beginning in the 19th century with crude oil processing, the products of which were often named differently in the various countries, resulted in a confusion of the terms existing up to now. This confusion has been increased further by the production of materials with similar characteristics, but of different origin, such as tar and peat. Thus the petroleum origin of asphalt and its correct designation, respectively, are not always familiar even to an expert, and erroneous conclusions are often drawn.

In the last decade efforts were made to develop a uniform nomenclature for bituminous materials all over the world. Although no uniform agreement could be reached in this field in spite of repeated international conferences, the application of a correct and suitable nomenclature was promoted by them[4].

Bitumen is the residue of certain crude oils on removal of the volatile components. The concentration process may take place in nature, or result artificially. The formation of natural bitumen and that produced by the distillation of crude oil are similar, but the bitumen obtained by distillation is produced more quickly and at a higher temperature.

The natural and artificial mixtures of bitumen and mineral matter are called asphalts. In general, natural asphalts were formed by the volatile components of crude oils having evaporated upon nearing the ground, and the residue seeping through the splits of the porous rock and being mixed with mineral matter. The mixtures formed are known as "natural asphalts". The term "asphalt" is used also for artificial mixtures made of bitumen and mineral matter. Such asphalts are sand asphalt, asphalt concrete, sheet asphalt, asphalt Macadam, asphalt mastic. The term "asphaltic bitumen" is also used in certain countries to denote bitumen free from mineral matter. The relationship between bitumen and asphalt is similar to that between cement and concrete.

As can be seen from the above statements, the terms "asphalt" and "bitumen" or "asphaltic bitumen" refers to natural substances of petroleum origin and artificial products derived from them. The residues obtained by distillation of wood, coal, and brown coal, i.e. pitch, are called neither asphalt nor bitumen due to their different origin, although they are included into bituminous materials in the literature. Under the heading of "raw materials", they are referred to as tars, tar pitches, or pitches. It should be mentioned that this nomenclature does not influence the terms in use for analytical investigations, as hard and soft asphalt, asphaltenes.

As has been stated in the foregoing concerning the individual terms, no distinction could be reached nor could uniform terms be accepted all over the world. Therefore it seems to be necessary to quote the designation of bitumen and asphalt of the American Society for Testing Materials (ASTM Designation D 8)[(5)].

Bitumen-Hydrocarbon material of natural or pyrogeneous origin, or combination of both, frequently accompanied by their nonmetallic derivatives, which may be gaseous, liquid, semisolid, or solid, and which is completely soluble in carbon disulfide.

Asphalt, a dark brown to black cementitious material, solid or semisolid in consistency, in which the predominating constituents are bitumens which occur in nature as such or are obtained as residue in refining petroleum.

In the United States literature the word "asphalt" is largely used, whereas the European and other references mostly are to "bitumen" or "asphaltic bitumen". According to the U.S. practice, the term "asphalt" will be used subsequently both for the natural product and that obtained from crude oil, as well as for products containing mineral matter also.

1.2 Natural Occurrence

All stages from the asphalt which is almost free from mineral matter to the rock with small asphalt content can be found in nature. The mineral matter quantity and the hardness of the asphalt is considered in the classification of natural asphalts. The origin is usually the basis of a commercial classification.[6,7]

The natural asphalt best known is the asphalt obtained from the Trinidad Lake[8]. The surface of the lake is 100 acres, its depth is more than 200 ft. The natural "lake asphalt" contains 39% bitumen, 30% mineral matter (partly in a colloidal state), and 31% emulsified salt water. The finished product obtained by purification is the refined Trinidad asphalt, or "Trinidad Epuré", a product with stable characteristics. The part soluble in carbon disulfide amounts to 56–57%[9]. The Bermudez Lake is bigger than the Trinidad Lake, but its average depth is only 8 feet.

The asphalt content of natural Bermudez asphalt is 64%, the mineral matter quantity amounts to 2%, and the water content is 30%. Besides, it contains 4% non-mineral components insoluble in carbon disulfide. The part of the refined product soluble in carbon disulfide is about 95%.

Asphaltites are hard natural bitumens containing in general almost no mineral matter. The best known asphaltites are Gilsonite and Grahamite. Also the Glance Pitches belong to this group. Their characteristics are between those of the above mentioned two materials. Gilsonite, which is soluble in carbon disulfide up to 98%, is one of the purest natural products.

The asphalt rocks (Asphalt Limestones and Asphalt Sands) characterized by their high mineral matter contents, also belong to the group of natural asphalts. Natural asphalt sands occur also in a lot of areas in various European countries. A characteristic example for asphalt formation from crude is represented by the occurrence of natural sand asphalts in Rumania, near the Hungarian frontier. In the opinion of Nyul[10] this material should be considered a concentrated crude oil.

The sand surfaces soaked with crude oil can be followed up in places deprived of their surface layers from the village of Tataros through Felsöderna up to Bodonos. The formation is inclined towards the middle of the Great Hungarian Plane and has been investigated to a great extent by drilling of slight depth.

It has been stated that the raw material in this area is still subject to the asphalt formation process. The laboratory test of the oil sand resulted in the characteristic data of a highly viscous crude oil, with high density, high softening point, and relatively low flash point. Besides the residual asphalt, fractions are formed in the laboratory distillation of the crude oil similar to those obtained from the naphthene base crudes of Rumania and Texas.

To process to asphalt the raw materials obtained by open cut method or in simple shafts, the sandy parts and the lighter oily parts are removed with hot caustic wash, by sedimentation, and, if possible, by subsequent vacuum distillation.

On reviewing the various asphalt occurrences not enumerated here, it can be stated that they are found in the very same areas where the large mineral oil deposits lie.

1.3 Chemical Constitution of Crude Oils

Crude oil quality is of first importance for the characteristics of the asphalt to be produced from it.

Crude oil composition is not uniform and it may be stated that there exist nearly as many kinds of crude oils as the number of oil fields all over the world. As for its composition, crude oil is a mixture of numerous hydrocarbon components, paraffinic or saturated, olefinic or unsaturated, cycloparaffinic or naphthenic, and aromatic hydrocarbons(11,12).

The least reactive hydrocarbons are the paraffinic compounds. Olefins do not occur generally in crude oils, and are the result of cracking only. They are the most reactive hydrocarbons. Naphthenes are saturated hydrocarbons too, but do not consist of straight or branched chains like paraffins. They contain rings with various numbers of carbon atoms. As a consequence of their saturated characteristics, the reactivity of the naphthenic hydrocarbons is very slight. However, they can be attacked more easily than pure paraffins. There are various benzene derivatives among the aromatics. They are more reactive than paraffins or naphthenes. The molecular constitution is much more intricate in higher boiling crude oil fractions and residues(13). These

TABLE 1.
Elementary Composition of Crudes

Element	%
C	81—87
H	10—14
S	0—6
O	0—0.7
N	0—1.2

hydrocarbons may contain naphthenic and aromatic rings, as well as paraffinic chains. In general, the higher boiling a distillate, the greater the cyclic character of the compounds. Also the carbon/hydrogen ratio is shifted more and more in favour of the carbon.

Beside hydrocarbon compounds, oxygen, nitrogen, and sulfur compounds occur in crude oils. Only very few of the hydrocarbons present in crude oils could be isolated in a pure state up to now. The reason for this is that these compounds are very much alike both chemically and physically, and their resolution is thus rather difficult[14]. Average data for crude oil elemental constitutions are given in Table 1. Almost every metal could be detected in the ashes of petroleum residues, if only in traces. Investigating the asphalt content of crude oils in detail, some researchers start from the crude itself. As for the chemical nature of asphalt constituents, Winniford and Witherspoon[15] state that a solubility series of asphalt constituents may be arranged according to their separation possibilities. Asphaltic materials are present in the crude oil as a colloid phase.

1.4 Asphalt Constitution

Depending on the constitution of the individual crude oils, the asphalts obtained from them contain various hydrocarbons and hydrocarbon groups in different proportions. This is influenced even in the case of a given crude oil by the asphalt production methods used. It is therefore obvious that no substantial results could be achieved up to now in the investigation of the individual compounds of asphalt and that research is being carried out by various methods. The procedure utilised most commonly in this area is the separation of asphalt in definite hydrocarbon groups and the study of their properties and distribution.

The group analysis methods used in most cases are based on the different solubilities of the various asphalt components in the individual solvents. With a great many methods this separation by solvent extraction is combined with selective adsorption.

Chromatographic adsorption procedures are also developed by some researchers. In recent methods the groups separated by solvent extraction and by selective adsorption are resolved in further characteristic groups by molecular distillation, fractional crystallization, chemical reactions, thermal diffusion etc.[16,17,18,19,20,21,22,23]

The following groups are determined in asphalt according to Richardson[24]:

Carboids	insoluble in carbon disulfide,
Carbenes	insoluble in carbon tetrachloride and soluble in carbon disulfide,
Asphaltenes	insoluble in low boiling saturated hydrocarbons, soluble in carbon tetrachloride,
Malthenes	soluble in low boiling saturated hydrocarbons.

The carboid and carbene contents of commercial asphalts are slight and increase only during cracking. Asphalt quality is determined by the quantity and nature of asphaltene and malthene components.

As for separation of asphalt groups by solvents, any number of other methods are developed isolating asphalt into further groups such as

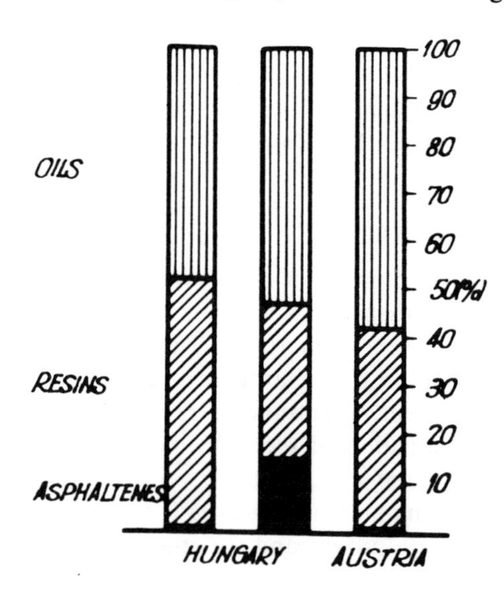

Fig. 1. Group composition of asphalts of different origin with nearly identical softening points (50°C)

hard asphalt, asphaltic resin, oil resin, and oily constituents[25]. In the well known and relatively simple procedures of group analysis, asphalts are separated into asphaltenes, resinous and oily constituents[26,27]. The physical properties and chemical compositions of these groups present a continuous transition according to their sequence, thus the limits are not sharp. The possible limits between the groups depend on separation conditions and on the method applied. It is therefore necessary to indicate the separation procedure together with the individual groups. Based on the investigations of Nyul, Zakar, and Mózes[28], data for various group compositions of asphalts of different origin and nearly identical softening points are shown in Figure 1.

To determine the group composition of the asphalts shown in Figure 1, test benzine, fuller's earth, and chloroform were used (method I). Zakar and Mózes[29] used n-hexane, benzene, and pyridine solvents, as well as silica gel adsorbent in their further tests (method II). The above mentioned asphalts could be separated into four groups by this method. The group composition data obtained by the two methods are compiled in Table 2 for comparison.

The above mentioned methods are suitable for the chemical and physical characterization of asphalt fractions and for the determination of their proportions. A relatively clear picture of the material in question is thus obtained. Valuable correlations may often be obtained in practice by the comparative utilisation of the given methods on asphalts of different origin and manufacture, and on asphalts of similar origin but different manufacture, respectively. Owing to the above mentioned intricate composition of asphalt, however, even these studies did not give the expected results.

It is worth mentioning that multinuclear compounds of acid cha-

TABLE 2.

Comparison of Methods to Determine Group Composition (Nagylengyel Asphalt, 85 Penetration, 49°C Softening Point)

Method of determination		*Composition, %*	
I	II	I	II
Asphaltene	Asphaltene	15	19
—	Asphaltic resin	—	10
Resinous comp.	—	32	—
—	Oil resin	—	26
Oily comp.	Oily comp.	53	45

racter were found in the oily constituents of asphalt by various researchers[30,31]. The acid compounds of the asphalt constituents are called asphaltic acids in the literature.

As a result of investigations, it was stated by Caro[32], that the appearance of the acids is different according to their origin.

As for elementary composition, asphalts contain 80 to 85% carbon and 9 to 10% hydrogen. Besides, there are different quantities of oxygen, sulfur, nitrogen in them, and in some cases also traces of halogens. The oxygen content is from 2 to 8% according to origin and production methods. Thus Caucasian asphalts possess a relatively high oxygen content. Nitrogen content is generally negligible, although the resinous parts of some California asphalts contain larger nitrogen quantities from 0.5 to 1.0%. The sulfur content of asphalts of various origin is between 0.5% and 7%. The sulfur contents of a few well known asphalts are shown in Table 3.

Asphalt contains also metallo-organic compounds. Among others, the metals occurring most frequently are iron, nickel, vanadium and calcium. The vanadium content of individual asphalt types is of outstanding interest since their ashes can be used to produce vanadium. According to Bor[33], the vanadium content of Nagylengyel crude oil amounts to about 0.008%, and that of the residue to approximately

TABLE 3.
Sulfur Content of Asphalts of Various Origin

Origin		*Sulfur content, %*
Argentine		0.6
Austria	Matzen	0.8
Hungary	Nagylengyel	5.0
	Barabásszeg	1.4
	Lispe	0.7
Mexico	Tampico	6.1
Near East		3.9
Rumania		0.7
Soviet Union	Baku	0.8
	Radajevka	3.0
	Romashkino	3.5
	Tuimaza	3.0
Texas		2.7
USA	Wyoming	4.6
	California (Edna)	3.2
	(Kern)	1.2
	Arkansas (Irma)	3.5
Venezuela		3.5

0.01%. It was found in agreement with other researchers that the vanadium content of crude oil is accumulated in the asphaltene containing constituents, that is, in the asphalt proper. The vanadium content of hard asphalt from a Nagylengyel asphalt of 55 penetration is approximately 0.03%.

1.5 Characteristic Properties of Asphalt

The characteristic valuable properties of asphalt are closely connected with its colloidal structure. There is a dispersion phase and a disperse phase finely distributed in it. The modern colloid chemical view on asphalt, as summarized by Nüssel[34], looks upon it as a colloidal two-phase system, chiefly consisting of a closed oily phase corresponding approximately to the malthenes, and of a solid phase dispersed in it. The oily phase is considered to consist of a great number of compounds, where each is soluble in the mixture of the rest. Only the highest molecular weight constituents, the so-called heavy asphaltenes are insoluble. They are dispersed in the oily medium and are surrounded due to adsorption phenomena with protective layers of the next lower molecular weight compounds corresponding approximately to the asphaltic resins. These represent a transition from the dispersed heavy asphaltenes towards the oily phase. Asphaltenes and protective layer together represent the micelle.

The colloidal structure of asphalt is proved by its rheological properties such as plasticity, elasticity, thixotropy. Colloidal constitution is indicated also when working with the electron microscope or by X-ray investigations. This conception is supported also by the Tyndall effect and Brownian movement which can be noticed at suitable solvent dilutions.

On the basis of the extent of micelle dissolution, asphalt can be classified as sol asphalts (fluid type) and gel asphalts (structural type) with the transitions between the two limiting cases consisting of various percentages of sol and gel structures[35].

The colloid constitution of asphalt depends in the first place on the following factors:

1. Chemical nature and percentage quantity of the asphaltenes,
2. Chemical character and percentage quantity of the malthenes,
3. Temperature of the system.

If the malthenes contain a sufficient amount of aromatics related to the quantity and quality of the asphaltenes, a sol structure will be formed, whereas with a low aromatic content gel structure will result.

Thus the colloidal constitution of asphalt does not depend only on the quantity of the asphaltenes.

Under the influence of temperature, the asphaltenes are more and more dissolved in the resinous constituent, while the latter is dissolved in the oily medium. Therefore the rheological properties of the oily phase prevail more and more. The gel structure is changed into sol structure by temperature increase. With decreasing temperature, the anomalous rheological properties begin to predominate. The colloidal constitution of the system determines the rheological properties of the asphalt to a great extent.

Asphalts are materials of high viscosity, almost solid at room temperature, and they become liquids when the temperature increases. These asphalt properties belong therefore to rheology the science dealing with flow phenomena, deformation, and changes in shape(36). It is known that liquids for which Newton's Law holds good are called Newtonian or normal liquids.

With normal liquids, the deformation stress in consequence ofthe force of gravity remains constant. This correlation is not valid for anomalous liquids, but it is always different. Most of the asphalts exhibit various properties according to their colloid structure belonging to anomalous liquids at room temperature. At higher temperatures, however, asphalt is also a Newtonian liquid.

The differences between the individual asphalts at surrounding temperatures are caused by their different rheological behavior, due to the sol or gel structure. The difference in characteristics of the individual asphalts at a given temperature depends on differences in Newtonian flow, the differences in hardness on the value of viscosity. Asphalt characteristics will differ when they are manufactured from the same crude by different methods, or from different crudes by the same method. Thus brittle, pitchlike asphalts or elastic asphalts somewhat like rubber can be produced having properties between the two limits(37).

This empirical classification was proved later by rheological investigations. The brittle, pitchlike asphalts are rheologically similar to pitch, as if they were Newtonian liquids. The gel structures undergo elastic deformation and show complete elastic recovery in the case of small changes in shape at room temperature. The sol structures possess intermediary characteristics between the two extreme asphalt groups from the rheological viewpoint. Based on these investigations, asphalts are distinguished as viscous sol or gel asphalts. It must be mentioned that the two extreme types can be distinguished exactly, but the separation of sol asphalt groups is not sharp enough.

Suitable test apparatus[38] is needed in the first place to carry out rheological investigations for the determination of asphalt characteristics and that of the absolute viscosity at ambient temperatures. For practical purposes, standardized short time tests are applied such as the determinations of softening point, penetration, ductility, breaking point. Each method is fundamentally a viscosity measurement carried out under arbitrary conditions to characterize some flow properties. To distinguish between the different types, the Pfeiffer and van Doormal[39] penetration index (P.I.) was introduced considering the temperature susceptibility. The individual asphalt types may be classified into three groups limited by certain arbitrary assumptions and knowing the softening point as well as the penetration at 25°C. The P.I. of pitch-like asphalts is smaller than −2, that of the gel asphalts larger than +2.

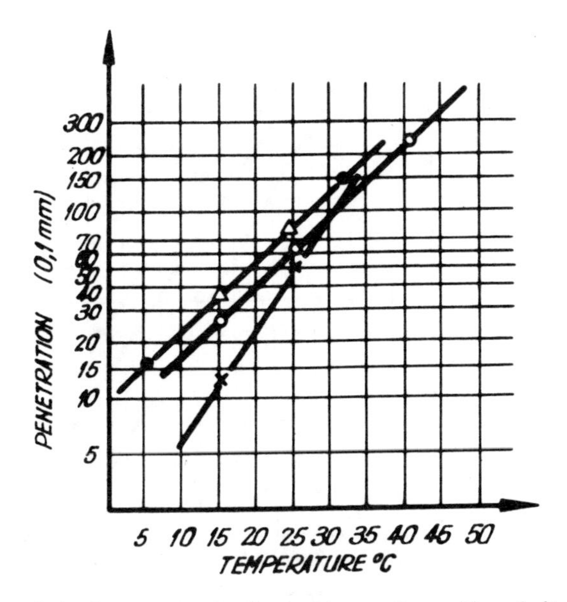

Fig. 2. Correlation between penetration and temperature with asphalts of various origin

● ● ●	Hungary (Nagylengyel)	S.p. 49°C
▲ ▲ ▲	Mexico	S.p. 49°C
○ ○ ○	Venezuela	S.p. 50°C
× × ×	California	S.p. 49°C

Beside the penetration index, there is also among other methods the determination of the so-called plasticity-temperature interval, which means the difference between the softening point and the breaking point of the asphalt expressed in °C. This difference increases constantly as from pitch to the gel asphalts.

The changes in asphalt consistency may be characterized at temperatures below the softening point best by the penetration measured at

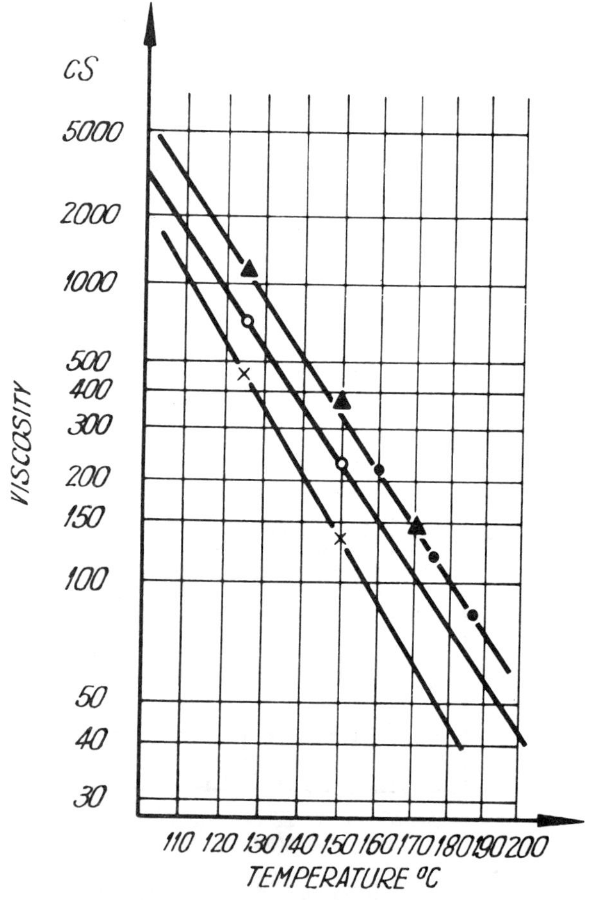

Fig. 3. Viscosity-temperature correlation with asphalts of different origin

● ● ●	Hungary (Nagylengyel)	S.p. 49°C
▲ ▲ ▲	Mexico	S.p. 49°C
○ ○ ○	Venezuela	S.p. 50°C
× × ×	California	S.p. 49°C

various temperatures and by the tangent representing the log penetration-temperature correlation, respectively[40]. The differences in the service behavior of various asphalts are shown suitably by the values and correlations thus obtained.

The change in penetration as a function of temperature is largest with Newtonian asphalt and smallest in the case of elastic gellike asphalts. The differences between asphalts of various origin are shown in Figure 2. The two extreme qualities of different character are California and Mexico asphalts, and the Nagylengyel asphalts, respectively[41].

The viscosity, measured at high temperatures, changes practically uniformly with all three groups. The differences in temperature susceptibility become smaller with temperature increase. Asphalts with similar penetration at 25°C exhibit different viscosities at high temperatures. The Newtonian asphalts belonging to Group 1 possess the smallest viscosities, those of Group 3 the largest viscosities. Figure 3 gives the viscosities of asphalts of various origin and similar softening points at high temperatures. The viscosity of the California asphalt is the lowest, whereas the highest viscosities are shown by the Mexico and Nagylengyel asphalts, respectively, similar to the different characteristics to be seen in Figure 2. It is to be mentioned here that a great number of researchers endeavour to develop a system to give a general characterization of the rheological properties of asphalts. Above all, the goal should be to obtain a universal picture by which the design and plant engineers would be able to summarize and calculate asphalt behavior quickly under service conditions. This rather intricate task has not yet been solved, since not only processing and service temperatures, but also service times and loads ought to be considered[42,43].

1.6 Physical and Chemical Characteristics

The thorough investigation of individual asphalt properties are treated in several special and comprehensive works[44]. Only such physical and chemical characteristics will be considered, as have not yet been treated and are most commonly used in practice.

The specific gravity of straight asphalt increases with decreasing penetration and increasing softening point, respectively. To illustrate this, the correlations between these data for Nagylengyel asphalt are compiled in Table 4[45]. The specific gravity of blown asphalt depends on the material used for blowing, but it is generally smaller than that of straight asphalts. The informative values from Table 5 may be used

TABLE 4.
Changes in Specific Gravity of Nagylengyel Asphalt with Progressing Distillation

Softening point °C	*Penetration at 25°C 0.1 mm*	*Specific gravity at 25°C g/ml*
44	162	1.026
51	76	1.038
65	30	1.044
74	17	1.050

TABLE 5.
Typical Specific Gravity Ranges for Various Asphalts

Penetration of asphalts at 25°C 0.1 mm	*Specific gravity at 25°C g/ml*
300	1.01 ± 0.02
200	1.02 ± 0.02
100	1.02 ± 0.02
50	1.03 ± 0.02
25	1.04 ± 0.02
15	1.04 ± 0.02
10	1.05 ± 0.02
5	1.07 ± 0.03
<5	1.07 ± 0.03

in general practice(46). The specific gravity of cracked asphalt is generally 0.10 higher than the data shown in the Table.

The practical importance of asphalt specific gravity consists in the possibility of converting masses into volumes and vice versa. Specific gravity decreases with increasing temperatures. The temperature coefficient of expansion for asphalt of various origin averages 0.0006 per °C from 15 to 200°C. By means of this, the given specific gravity determined at 25°C for asphalt in general may be converted to the temperature desired.

The specific heat (cal/gram grade) may be considered identical for the various asphalts. The value increases with temperature increase and is approximately half of the specific heat of water. Since asphalt possess no melting point, it has no heat of melting. On heating of asphalt, the following average values are applied:

at	0°C	0.4 cal/gram grade
″	100°C	0.45 ″ ″ ″
″	200°C	0.5 ″ ″ ″
″	300°C	0.55 ″ ″ ″

Thermal conductivity is practically identical with all asphalt types and decreases somewhat with increasing temperature.

Its value amounts to

0.14 kcal/m °C hr	at 0°C
0.13 kcal/m °C hr	at 70°C

It can be seen that the thermal conductivity of asphalt is very slight owing to which asphalts exert a pronounced heat insulating effect.

The surface tension of asphalt is 24.1 to 26.9 dyne/cm^2 at 150°C according to Saal. A range from 32.1 to 34.4 dyne/cm^2 can be calculated from this for 25°C.

Among the electrical properties of asphalt, the knowledge of the dielectric strength, the specific conductance, the dielectric losses and the dielectric constant is important.

The minimum value of the electric field intensity at which a disruptive discharge occurs, that is, the dielectric strength depends on production conditions and is from 10 to 60 kv/mm at 20°C. The dielectric strength of soft asphalts is smaller than that of hard asphalts. It decreases with increasing temperature in the same asphalt.

The value of specific conductance amounts to

$30. . .20 \times 10^{-13}$ ohm^{-1} cm^{-1} at 80°C

The conductance increases when the temperature rises. The following average values are used for asphalt in electrical engineering for the dielectric losses and the dielectric constant:

Tangent of total loss angle × 100	at 20°C	=1.3—2.1
Hysteresis losses and power losses	at 80°C	=3—5
Dielectric constant	at 80°C	=2.9—3.2

It can be summarized that asphalts possess good electrical insulating properties and are in no way inferior to other good dielectrics (insulators).

A very important property of asphalt is its low permeability to water. Its value viz.

$1—2,\ 4 \times 10^{-8}$ g/h cm Torr

is considerably lower than that of caoutchouc and of many plastics. The water absorption of asphalts is very slight and does not exceed 1—3%, even after long water storage, depending on asphalt hardness.

Asphalt is highly resistant to a number of chemical agents at ambient temperatures. It reacts with certain substances such as oxygen, sulfur, chlorine at higher temperatures. This property is applied in the manufacture of various asphalt types. Mainly dehydrogenation and asphaltene formation take place during these processes.

Chemical changes are experienced in asphalts at high temperatures even without the presence of outside material. Asphaltenes are formed during long heating from 300 to 350°C. At still higher temperatures carbenes and carboids will result.

Reactivity and heat resistance of asphalts depend on their chemical composition. Owing to this, differences are experienced when asphalt types of different origin and manufacture are investigated by means of various methods. Relationship between composition and quality on the one hand, and oxydation resistance on the other hand is not yet clear. More exact data from this viewpoint have only been obtained with cracked products up to now.

Asphalt properties undergo a slow change under the influence of air and light, aging occurs. The extent of oxydation depends on the specific surface exposed to air, and on the diffusion rate of the latter. Asphalt oxydation is slow in the absence of light at surrounding temperatures. It is accelerated by the influence of light. Insoluble, brown coloured condensation products are formed, and the asphalt surface becomes dull and brown. Besides, water soluble, acidic decomposition products result[47].

Asphalt as a protective material is often applied in lagers several millimeters thick. As shown by tests, the effect of air and light on these layers is limited to the surface only, asphalt being therefore very suitable as a protective material. Usually thin layers are formed when asphalt is used as a bonding material. Light cannot reach the asphalt in this case. Asphalts are only changed by oxydation if they contact air.

Asphalts possess a great many advantages for the protection against the attack of acids, of carbonic acid, and water solutions of inorganic salts[48]. The resistance of asphalt against acids depends on the concentration of the latter. Dilute acids do not react with asphalt, but it is usually attacked by concentrated acids. However, it is very resistant against concentrated hydrochloric acid. Some types of asphalts, especially soft asphalts with a high acid value are attacked by dilute bases. This is not the case with concentrated sodium hydroxide (20%) and sodium carbonate (10%) solutions at room temperature.

Working on Romashkino asphalt, Gundermann and Kloss[49] stated that it was resistant against acids and bases. The amount of corrosion depends on the hardness, the origin, and the previous treatment. Asphalt is dissolved by most organic solvents with the exception of low molecular weight alcohols, as well as of oils and greases. The harder the asphalt, the higher its resistance against chemical action.

1.7 Test Methods

Several tests have been developed for the quality evaluations of the asphalts manufactured. Test methods are standardized in the individual countries[50,51,52]. The methods used commonly for evaluation are practically identical, with negligible differences; the comparison of literature data presents therefore no difficulties. Other quality tests are only partly corresponding to standards and in many cases are different in each country. When using a general term, as paraffin wax content, the method used should be indicated. There are still greater differences in the special tests developed by various researchers.

The aim of the methods generally used is the determination of asphalt consistency, degree of purity, and heat stability. Several suitable methods were developed to determine consistency.

They furnish data connected more or less with the viscosity. Asphalts are characterized and/or compared either by the flow experienced at a given temperature, or by the temperature belonging to certain properties. The tests to be described, penetration, softening point, ductility, breaking point are not equivalent to direct viscosity determinations, but are commonly used in the industry, for the purpose of determining asphalt consistency rapidly.

1.7.1 Penetration

Penetration is a test for the consistency of asphalt. It is based on a standard needle entering asphalt under fixed conditions. The depth of the penetration of the standard needle into the material to be tested within 5 seconds, with a load of 100 g at 25 °C is measured in standard

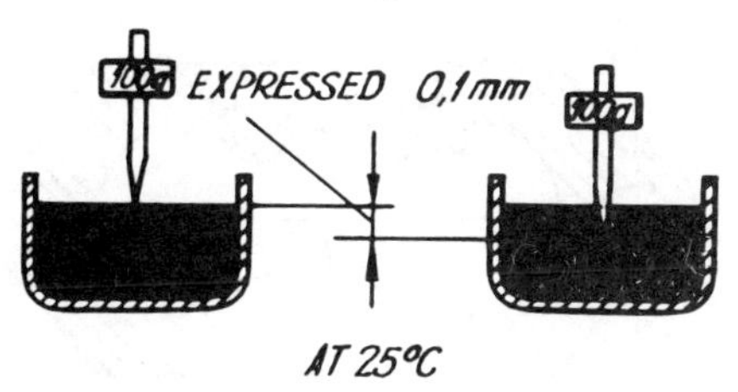

Fig. 4. Determination of penetration

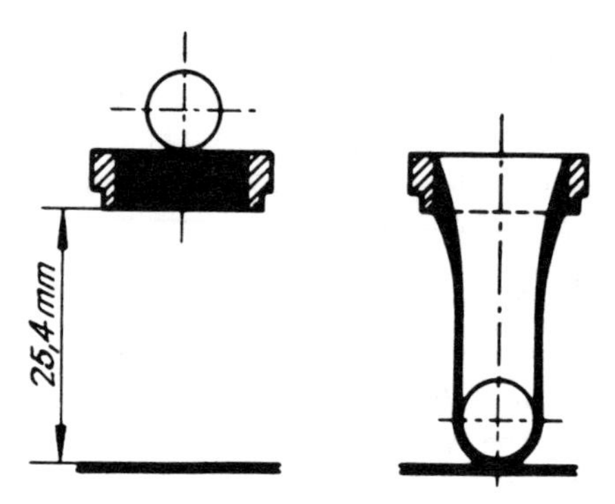

Fig. 5. Determination of softening point, Ring and Ball

tests (Figure 4). The penetration is measured in 0.1 mm, and the numerical value obtained is the asphalt penetration proper.

1.7.2 Softening Point

Asphalts show no melting points, since the transition from solid to liquid state does not occur at a definite temperature. Asphalt does not melt, not being a solid, but a highly viscous liquid like glass. Asphalt softens gradually on heating. The softening temperature determined under exactly described arbitrary conditions is thus called softening point. The softening point Ring and Ball Test is the temperature at which the asphalt layer placed in a standard ring touches the plate under the ring due to the weight of a ball of definite mass and size. (Figure 5).

1.7.3 Ductility

Asphalt ductility is the numerical value in centimeters, recorded as the length upon breaking of the asphalt thread formed under standard conditions of pulling.

Ductility determinations are carried out by ductilometers. The apparatus consists of a water bath in which at least three asphalt bodies

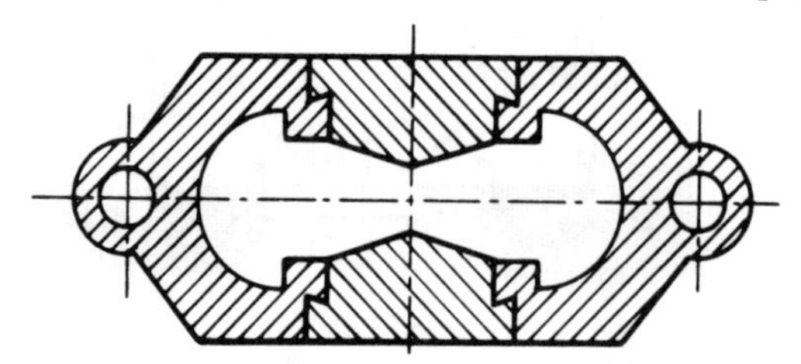

Fig. 6. Specified brass mould for asphalt to determine the ductility

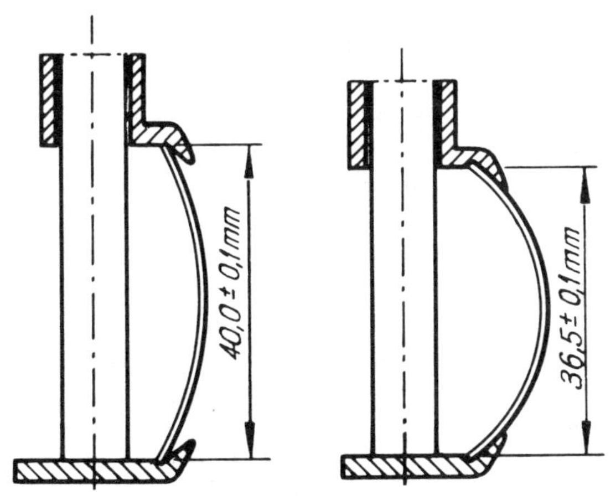

Fig. 7. Clamp for bending of a steel plate coated with asphalt for the determination of Fraass breaking point
a. Original setting b. Maximum bending

poured in standard moulds can be moved at a rate of 5 cm/min by means of a suitable device. The length of the asphalt thread formed by stretching is read on a measuring lathe. The test must always be carried out at a determined temperature. The test made at 25°C is generally given in the specifications. (Figure 6)

1.7.4 Determination of the Breaking Point

The Fraass breaking point is the temperature at which an asphalt layer placed on a standard steel plate will break when bent repeatedly in the same direction under fixed conditions. Based on the above principle, the bending device consists of two concentrated tubes at the lower end of which steel jaws are fixed. By the movement of the inner tube, a steel plate (Figure 7) fixed between the steel jaws is bent. The bending device is placed in a glass tube equipped with cooling.

1.7.5 Further Tests

Purity of asphalt is determined by its solubility in the prescribed solvents, by the ash content, the water soluble asphalt part, and the flash point.

Determination of loss on heating (volatility) informs on the heat stability of asphalt. Loss on heating of asphalt according to the standard represents the loss of weight after heating the material in question at 163°C for 5 hours in a standard capsule under fixed conditions.

This method is carried out in different countries by various methods, thereby yielding different results due not only to the differences in the material tested. Beside these tests, the paraffin wax content of the asphalt is often determined. According to the most commonly used method, the paraffin wax value of asphalt is the quantity of paraffin hydrocarbons which can be frozen from the oily component of asphalt under specified conditions at −20°C, and obtained by subsequent analytical separation. Although the asphalt viscosity is given as a specification only in some countries, the determination and knowledge of viscosity are necessary both for the various production processes, and for application procedures. Various viscosimeters can be used in principle to determine the viscosity at higher temperatures, where asphalt behavior is that of a Newtonian liquid, for example, capillary-, rotational-, and falling body viscosimeters, and technical viscosimeters. First of all, the technical viscosimeters such as Engler-, Saybolt-, Furol-, Redwood viscosimeters are used to determine pumpability, blending, and spraying qualities. They are commonly used in the petroleum industry due to their ease of handling, and they furnish data of adequate accuracy for the above purposes[53].

The drawback to these apparatuses consists in their not having been developed for high enough temperatures for asphalt tests. Up to now

TABLE 6.
Viscosity Conversion Factors

Known Viscosity	*Required Viscosity*							
	c.s. V_K	*°Engler*	*secs. Redw. I*	*secs. Redw. II*	*secs. S.T.V. 10 mm.*	*secs. S.T.V. 4 mm.*	*secs. Saybolt Univ.*	*secs. Saybolt Furol*
c.s. V_K	1	0.132	4.10	0.41	0.0025	0.076	4.7	0.47
°Engler	7.58	1	31.1	3.11	0.019	0.576	35.63	3.563
secs. Redwood I	0.244	0.0322	1	0.1	0.00061	0.0185	1.12	0.112
secs. Redwood II	2.44	0.322	10	1	0.0061	0.185	11.2	1.12
secs. S.T.V. 10 mm.	400	52.8	1640	164	1	30.4	1880	188.0
secs. S.T.V. 4 mm.	13.2	1.74	54.1	5.41	0.033	1	62.04	6.204
secs. Saybolt Univ.	0.213	0.028	0.873	0.0873	0.00053	0.0162	1	0.1
secs. Saybolt Furol	2.13	0.28	8.73	0.873	0.0053	0.162	10	1

no apparatus has been made for this purpose which is accepted generally. The values obtained with the above mentioned instruments makes is possible to state kinematic viscosities, whereas only practical numerical values such as °E, sec Saybolt, Furol, are furnished by technical viscosimeters. On the basis of literature data(54), the use of viscosity conversion tables is widespread to correlate the values obtained when working with various instruments (Table 6).

To determine the rheological properties of asphalts at the softening point, special viscosimeters become necessary for research purposes. Measurement conditions ought to be chosen corresponding to the nature of the material in question. Such instruments are the various rheoviscosimeters, consistometers, rotational viscosimeters(55).

Apart from the above test methods, further special test specifications are found in various standards. Thus the determination of asphalt constituents insoluble in cyclohexane serves the purpose of detecting a possible cracking of paving asphalts. A corresponding procedure was developed by Maas(56) in his earlier comparative investigations. This method was later incorporated in DIN 1995 based on the studies of Krenkler(57).

As for the accuracy of this test, justified objections were raised and the test has still to undergo another revision(58). The Oliensis spot test is a wide spread test but has not yet been commonly accepted. It must be stated that beside the tests for classification of asphalt and for the determination of its purity, tests for a uniform quality evaluation are lacking for asphalt, due to the intricate character of the problem. Various tests are in use beside the standard tests to correspond to the individual requirements of the field of application.

The U.S. specifications covering asphalt tests have been developed by the American Society for Testing Materials (ASTM). The American Association of State Highway Officials (AASHO) publishes similar tests for road building materials, generally following the ASTM Standards(59). Besides, the U.S. General Services Administration publishes federal specifications for bituminous materials, which are similar to ASTM Standards.
In addition to the U.S. specifications, the Institute of Petroleum (IP) London, England, publishes methods of testing for materials derived from petroleum(60).

The number of the test method to be used is always indicated in the quality specifications relating to individual asphalt types. These are dealt with in Chapter 3. For the sake of a clear survey, the methods relating to quality requirements of asphaltic materials have been summarized in a Table included into the Appendix. According to the scope of the book, no methods for testing paving and asphalt products are shown. However, there exist separate methods also relating to these materials.

References

1. Abraham, H.: Asphalts and allied Substances. New York (1960). D. Van Nostrand.
2. Carp, H.: Bitumen, *22*, 13 (1960).
3. Bitumen, Teere, Asphalte, Peche. *13*, 571 (1962).
4. Pfeiffer, J.: The properties of asphaltic Bitumen. New York (1950). Elsevier.
5. 1967 Book of ASTM Standars, Part 11 ASTM Philadelphia (1967).
6. Ullmann: Enzyklopädie der Technischen Chemie, Bd. 4. München (1953). Urban Schwarzenberg.
7. Kirk, R. E., u. Othmer, D. F.: Encyklopedia of Chemical Technology. New York (1948). Interscience.
8. Anon: Mining J. 259, 613 (1962).
9. Becker, W.: Strasse u. Autobahn. *13*, 131 (1962).
10. Nyul, Gy.: Magyar Mérnōk és Épitész-Egylet Kōzlōnye 75 köt 17-18 sz. (1941).
11. Sachanen, A. N.: The Chemical Constituents of Petroleum. New York (1945). Reinhold.
12. Lehmann, G.: Erdöldestillation. Mainz (1956).
13. Prinzler, H.: Einführung in die Technologie des Erdöls. Leipzig (1961).
14. Aixinger, I., a.c.: Ásványolajtechnológia. Budapest (1951). Nehézipari.
15. Winniford R. S., Witherspoon P. A.: Conference on the Chemistry and Chemical Processing of Petroleum and Natural Gas. Akadémiai Kiadó Budapest (1968). Sec. IV (1).
16. Grader, R.: Öl u. Kohle, *38*, 867 (1942).
17. Krenkler, K.: Bitumen, Teere, Asphalte, Peche. *2*, 59, 85, 105 (1951).
18. Bestougeff: Proc. IIIrd W. Petr. Congr. The Hague VI. 116 (1951).
19. O'Donnel: Analyt. Chem. *23*, 894 (1951).
20. Nyul, Gy., Mózes, Gy., Zakar, P.: Máfki 132 kiadv. Vesprém (1957).

21. Prinzler, H., Klimke, R., Drow, H.: Chem. Techn. *15*, 170 (1963).
22. Leibnitz, E., Hrapia, H., Papp. J.: Chem. Techn. *16*, 737 (1964).
23. Vajta, L., Vajta, S.: Erdöl-Kohle-Erdgas-Petrochem. *18*, 629 (1965).
24. Richardson, C.I.: The Modern Asphalt Pavement. New York J. Wiley & Sons (1905).
25. Pöll, H.: Erdöl und Teer, *8*, 350 (1932).
26. Strieter, C. G.: J. Res. nat. Bur. Standards. *26*, 415 (1941).
27. Hoiberg, A. J., Garris, W. E.: Ind. Engng. Chem. analyt. Edit. *16*, 294 (1944).
28. Nyul, Gy., Zakar, P. Mózes, Gy.: Erdöl u. Kohle, *12*, 967 (1959).
29. Zakar, P., Mózes, Gy.: Máfki 270. kiadv. Veszprém (1960).
30. Marcusson, J.: Z. angew. Chem. *29*, 21 (1916).
31. Bitumen, Teere, Asphalte, Peche. *13*, 526 (1962).
32. Caro, J. H.: Erdöl-Z. *78*, 342 (1962).
33. Bor, Gy.: Máfki 58. kiadv. Veszprém (1954).
34. Nüssel, H.: Bitumen. Mainz (1958). Hüthig und Dreyer.
35. Nüssel, H.: Bitumen. *17*, 204 (1955).
36. Mózes, Gy.: Máfki 158. kiadv. Veszprém (1958).
37. Zakar, P., Mózes, Gy.: Bitumen, Teere, Asphalte, Peche, *9*, 275 (1958)
38. Benes, V.: Asfalty, Praha (1961). Stat. Naklad. Tech. Lit.
39. Pfeiffer, J. Ph., Van Doormal, P. M.: J. Instn. Petroleum Technologists, *22*, 414 (1936).
40. Bitumen, Teere, Asphalte, Peche. *13*, 126 (1962).
41. Zakar., P. Mózes, Gy.: Bitumen, Teere, Asphalte, Peche. *13*, 248 (1962).
42. Van Der Poel, C.: J. appl. Chem. May 4. 221 (1954).
43. Saal, R. N. J.: Proc. Fourt World Petr. Congr. Rome (1965) Sec. VI/A, Prepr. 3.
44. Hoiberg, A. J.: Bituminous materials: Asphalts, Tars and Pitches Vol. I-II Interscience Publ. New York (1965).
45. Simon, M., Zakar, P.: Máfki 57. kiadv. Veszprém (1953).
46. Saal, R. N., Hevkelom, W., Blokker, P. C.: J. Inst. Petroleum, *26*, 29 (1940).
47. Gundermann, E.: Chem. Techn. *11*, 441 (1959).
48. Walter, H.: Bituminöse Stoffe im Bauwesen. Heidelberg (1962). Strassenbau.
49. Gundermann, E., Kloss, B.: Chem. Techn. *16*, 41 (1964).
50. Tekhnicseskie normy na Nefteprodukty Moscow (1957). Gostopt.

51. DIN 1995, Berlin 1960.
52. ASTM Standards on Bituminous Materials. Philadelphia (1960).
53. Vámos, E.: Ásványolajtermékek viszkozitása. Budapest (1951). Nehézipari.
54. Bitumen und Asphalt Taschenbuch, Wiesbaden (1957). ARBIT-Bauverlag.
55. Traxler, R.: Asphalt, New York (1961). Reinhald.
56. Maas. W. F.: Asphalt u. Teer, Strassenbautechn. *42*, 84 (1942).
57. Krenkler, K.: Bitumen, Teere, Asphalte, Peche, verwandte Stoffe *6*, 295 (1955).
58. Földes, E., Kádár, I.: Máfki Közlemények 8. sz. (1966).
59. Standard Specifications for Highway Materials and Methods of Sampling and Testing, AASHO Washington.
60. Standard Methods Testing of Petroleum Products, Institute of Petroleum London (1957).

2 Manufacture of Asphalt

2.1 Fundamentals

Most crude oils can be processed to obtain asphalt by selecting a suitable procedure. Since the quality and the asphalt content of crude oils are different, the manufacturing processes also differ considerably.

Asphalt can be obtained from crude oils having an asphalt content of from 5 to 10%. However, if asphalt production is the principal purpose, it is of course more advantageous to utilize crude oils with higher asphalt contents, amounting to at least 25%. In quite a few crude oils the asphalt content exceeds 50%, in some of them even 70%. Relatively few of such crude oils a are used for asphalt manufacture in countries where such crude oils prevail, due to the limited marketability, and most asphaltic crude oils is used as fuel oils, or submitted to the thermal cracking process. If asphaltic crude oil is not available, and no such crudes can be imported for asphalt manufacture, the asphalt requirements of the country must be ensured from the crude oil available. Sometimes several crude oils are available from which asphalt can be made. In such cases the most suitable manufacturing process must be selected considering the desired asphalt quality and the intended application. Beside crude oils suitable for making paving asphalts, asphaltic residues obtained during lubricating oil refining can be used for certain industrial purposes. Based on these considerations, the present international situation was developed. This is characterized by the fact that from most industrially processed crude oils, asphalt is recovered along with other products. Thus world asphalt production is based less and less on asphaltic crude oils such as Mexico and Venezuela crude oils, which represented raw materials of outstanding importance for asphalt manufacture only a few decades ago.

Asphalt manufacturing techniques from crude oils may be divided fundamentally into two groups:

1. Separation of asphalts from crude oils with sufficient asphalt

contents. This is effected by distillation or separation by selective solvents,

2. Production of asphalt from crude oils in which asphalt constituents are not present in sufficient quantities and additional amounts of these must be produced.

Considering these facts, the most important processes for asphalt manufacture are the following:

1. Distillation
2. Extraction
3. Blowing

The asphalts made by these processes are called straight run, propane, and blown asphalts.

It should be mentioned that no sharp separation of these methods can be made in asphalt manufacture in general practice. The formation of asphaltic material must also be considered as a result of chemical processes taking place during concentration when distillation is carried out under given circumstances. In the first place, temperature and heating time are decisive, since unsaturated compounds result by cracking of hydrocarbons and their derivatives. This in turn involves polymeriation to give high molecular weight malthenes and asphaltenes. The individual methods are also applied alternatively and subsequently, according to the given raw material and production conditions. Commercial asphalt is also made by distillation or extracting distillation residues, by blowing propane asphalt, blowing straight asphalts, and even by repeating or modifying the processes. In many cases it is suitable to process crude oils and various raw materials together, to apply additives. By blending together asphalts made by different manufacturing techniques or asphalts with other petroleum products, desired properties are obtained.

Asphalt manufacturing by cracking should also be mentioned. Opinions differ as to this being considered a separate method. There exists a possibility of producing asphalt by direct cracking, but most asphalt designated as cracked asphalt is made by further processing of crack residues, by distillation, blowing and/or blending.

Cracked asphalts are in general variable in their properties and are to be controlled more accurately for application, due to which a separate treatment of this process seems to be advisable. The most important manufacturing processes and their fundamental relationships are shown in Figure 8. The various blending possibbilities are not included. Only blending of propane asphalt with lubricating oil

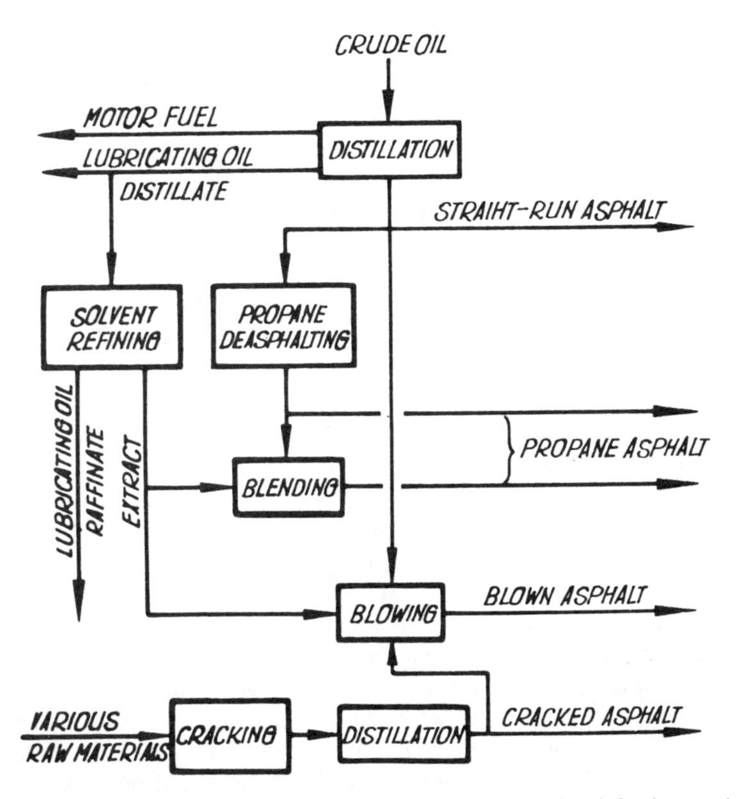

Fig. 8. The most important methods of asphalt manufacture and their fundamental relationships

extract is indicated. It ought to be mentioned again that combined methods are to be used in commercial manufacture according to the raw material and industrial circumstances, as well as to quality requirements.

As it has been mentioned, the bulk of commercial asphalts is produced from crude oils of various origins and having different asphalt contents. The given crude oil must therefore be investigated thoroughly, and the quality characteristics of the asphalt obtainable from the crude oil have to be determined by previous tests. After laboratory tests, pilot plant production of straight run asphalt is required, in suitable quantities and different qualities to enable comparison with the standards, and specifications of different countries. The straight asphalts or vacuum residues sometimes happen to correspond directly to the requirements. With some crude oils, however, further tests on the asphalt obtained from them becomes necessary due to the differences in softening point-penetration, breaking point, or paraffin wax content. In such cases techniques must be employed other than distillation, such as further blowing of residues of a given consistency, and the blending of these residues with suitably selected components. It often seems necessary to resort to plant assays on a large scale after preliminary tests, and to study products together with consumers, if necessary. The suitability of the given crude oil for asphalt manufacture may be decided finally on the basis of such test series.

2.2 Investigation of Crude Oils

To evaluate the various crude types in practice, paraffinic base and naphthenic (asphaltic) base crude oils are distinguished, depending on whether straight chain hydrocarbons (paraffins) or cyclic (aromatic or naphthenic) hydrocarbons prevail in the crude oil. Originally the paraffin wax content (melting point) and residue quality (asphalt) were considered decisive. This simplification does not meet the requirements, since intermediate base crude oils were also found, which include the crude oils of a great many oil fields. Furthermore, crude oil types with light distillates belonging to a group different from those yielding heavy distillates were also detected. Thus a relatively simple crude oil evaluation method had to be developed to test all the crude oils uniformly. Characteristic test data, calculated quantities based on experience, or structure designations agreed upon are applied. It was found that the crude test was valid when based on the assumption that the specific gravity and boiling point of open chain .hydrocarbons are lower than those of cyclic hydrocarbons and that the specific gravity of distillate mixtures having identical boiling points decreases with, the number of open chain linkages (paraffins).

According to this fundamental principle and upon thorough testing of several thousands of crude oils, a crude oil test method based on the so-called Hempel distillation was standardized by the Bureau of Mines (U.S.) This test method rests on the specific gravity of empirically selected fractions, termed the "key fractions". It indicate to which of the groups the distilled crude oil sample should be assigned based on its general chemical composition. The method consists in the distillation of the crude oil to be tested in a specified apparatus under standard conditions first at atmospheric pressure up to 275°C, then at

40 mm absolute pressure up to 300°C. Individual cuts are collected at every increase of 25°C, and the specific gravity of the last atmospheric and vacuum fraction, respectively, is determined at 60°F. Seven crude types are distinguished by the system between the limiting properties of "paraffin-paraffin" and "naphthene-naphthene"[1,2]. On the basis of work by Kerényi, Zakar, and Mózes[3] characteristic data of crude oils of various origins, as well as quality designations obtained by the above method are shown in Table 7. Comparison of the prescribed data to those received in the tests allows certain predictions. On comparison of the crudes from Barabásszeg and Szolnok, respec-

TABLE 7.
Test—and Quality Data for Crudes of Various Origin

Crude	*Spec. gravity at20°C, d 20$_4$*	*Viscosity (°E) 50°C*	*100°C*	*Pour point °C*	*Conradson carbon residue %*	*Sulfur %*	*Quality at atmospheric distillation*	*vacuum distillation*
Anastasievka (Soviet Union)	0.8960	1.6	1.2	−50	2.0	0.25	intermediate	intermediate
Barabásszeg (Hungary)	0.9333	—	45	+35	12.8	1.31	paraffin	paraffin
Buzsák (Hungary)	0.9690	25	4.6	−18	8.3	2.0	naphtene	naphtene
Lovászi (Hungary)	0.8781	—	—	—	0.3	0.06	intermediate	intermediate
Matzen (Austria)	0.9064	8.0	2.5	−53	2.6	0.20	intermediate	intermediate
Nagylengyel (Hungary)	0.9584	60	5.8	−4	13.7	3.82	paraffin	intermediate
Romashkino (Soviet Union)	0.8664	1.7	1.3	−48	5.4	1.80	intermediate	intermediate
Szolnok	0.8914	3.5	1.6	+37	4.3	0.59	paraffin	paraffin
Schönkirchen (Austria)	0.8586	4.8	1.6	−30	4.4	0.26	naphtene	naphtene
Tyulenovo (Bulgaria)	0.9364	12.4	2.2	−24	3.6	0.30	naphtene	intermediate
Tuimaza (Soviet Union)	0.9370	1.5	1.2	−40	4.2	1.46	paraffin	intermediate

tively, both paraffin-paraffin crude oils, the high Conradson carbon residue of the first crude oil is remarkable, indicating high asphalt content. With Nagylengyel and Tuimaza crude oils (both intermediate-paraffin) higher values of specific gravity, viscosity, pour point, sulfur content, and especially the very high Conradson carbon residue are rather striking. Considerable difference between the asphalt contents to be expected are indicated by these values. The higher Conradson carbon residue and the naphthene-naphthene character of Buzsák crude oil point to good quality asphalt being obtained from it.

However, it is obvious that no suitable characteristics are given by this method for extensive evaluation, since the other crude oils included in the Table are paraffin-paraffin, paraffin-intermediate, and intermediate-intermediate types. Thus Lovászi and Matzen crude oils having different distillate compositions belong in the same group. At the same time Matzen, Schönkirchen, Tuimaza, Romashkino crude oils, as well as Tyulenovo- and Buzsák crude oils are included in another group because of very slight differences in specific gravity.

Recent characterization systems rest not only on the specific gravity, but also on other crude oil properties. The Soviet method[4] for crude oil evaluation is based on the selection and compilation of the most important technical features, such as qualifying quantity, classification, and type of the crude oil, as well as some other data. Low or high sulfur content, low, middle, or high resin content, low, middle or high paraffin wax content of the crude oil serve as the bases of its classification. In the type classification, the octane number of gasoline, and that of the petroleum distillate, as well as the viscosity index of lubricating oil distillate are predicted on the basis of agreed group limits in a similar way. Definite properties of the individual products are indicated by these characteristics. Due to the fundamental assumptions of this method, the group number of crude oils having better technological qualities becomes more often Class 1, and less frequently Class 3.

However, when the applicability of a crude oil or the development of the most suitable process techniques are to be dealt with, knowledge of characteristic data on distillate distribution and product quality must be available in practice beside this classification method, which is hardly sufficient for this purpose. Kerényi[5] intended to supply this information in his method of crude oil evaluation, based on the exact correlation between the distribution curves of molecular size and structure on the one hand, and physicochemical as well as technical properties on the other hand. To characterize crude oils, four constants des-

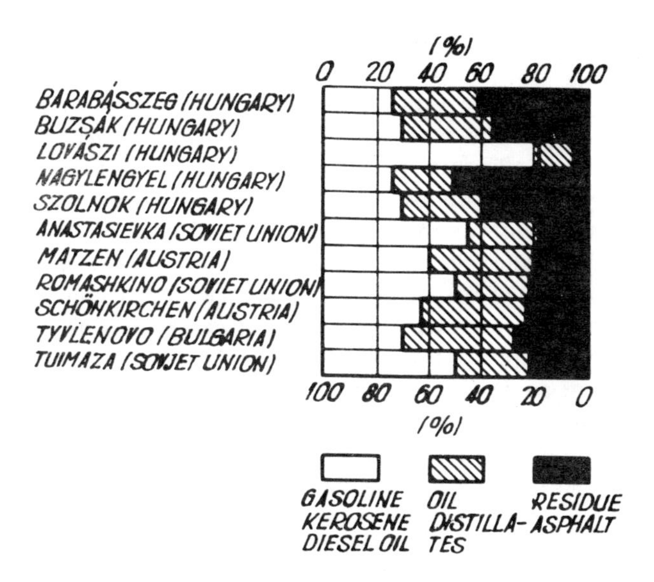

Fig. 9. Crude oil analysis

cribing distillate distribution are given, together with data on the aromatic and naphthene rings at the 10 and 50 vol. % points. Although this method covers the most important characteristics of distillate constitution, and product quality, no information on residues are obtained by it.

An approximate method is given by Nelson and Suresh Patel[6] to predict the asphalt contents of crudes as well as the asphalt quality on the basis of four test values.

Figure 9 shows the distillate distribution data of the crude oils enumerated in Table 7.

The yields shown in the Figure were determined by Hempel distillation used for comparison tests.

Beside temperature ranges for the individual products, vacuum distillation is carried out at a temperature corresponding to the atmospheric distillation temperature of 555°C (7).

The distillate composition of the crude oils given in Figure 9 will not be discussed here, since only residues are important in asphalt manufacture. Apart from the large residue quantity of Nagylengyel crude oil, the residue quantities of Barabásszeg, Buzsák, and Szolnok crude oils are very much alike. No substantial differences can be experienced with the crude oils from Anastasievka, Matzen, Romashkino, Schönkirchen, Tyulenovo, and Tuimaza. The lowest residue is obtained from the Lovászi crude oil. The data of Table 7 contain only hints regarding the quality of these residues. An exception is Buzsák crude with its naphthene-naphthene character, from which an asphalt of very low paraffin wax content would probably result by processing. Data on yields and groups of crude oils similar to each other are however not sufficient. In connection with the asphalt quantity and quality of the individual crude oils, the characteristic data of vacuum residues obtained in the above mentioned laboratory investigation should be compared. These are given in Table 8. It can be seen that there are substantial differences between the asphalt characteristics of residues from similar crude oils. Testing and previous laboratory investigation, respectively, of given crude oils by the above methods result in the most important information on certain questions within a wide range. Further studies however are required as to the application of a crude

TABLE 8.
Characteristics of Vacuum Residues from Various Crudes

Crude	*Specific gravity at 25°C g/cm³*	*Softening point °C*	*Penetration at 25°C 0.1 mm*	*Sulfur %*
Anastasievka (Soviet Union)	0.995	Pour point + 2929	non measurable	0.5
Barabásszeg (Hungary)	—	170	0	2.4
Buzsák (Hungary)	1.001	25	non measurable	0.60
Lovászi (Hungary)	—	—	—	0.38
Matzen (Austria)	—	—	—	0.48
Nagylengyel (Hungary)	1.071	78	17	5.6
Romashkino (Soviet Union)	—	46	67	2.7
Szolnok (Hungary)	—	55	136	0.74
Schönkirchen (Austria)	—	—	—	0.45
Tyulenovo (Bulgaria)	0.976	Pour point + 21	non measurable	0.37
Tuimaza (Soviet Union)	—	36	296	3.0

TABLE 9.
Crudes with High Asphalt Content

Crude	*Specific gravity at 15°C g/cm³*	*Sulfur %*	*Viscosity °E/37,8°C*	*Carbon residue %*	*Asphalt Yield (100 penetration at 25°C) %*
USA Wyoming, Oreg. Bas.	0.922	3.37	8.5	7.3	44.6
USA Arkansas, Irma	0.962	2.54	110	11.2	49.7
USA California, Kern River	0.972	0.93	170	7.5	54.6
Mexico, Tampico	0.993	5.15	170	14.4	73.6
Hungary, Nagylengyel (1960 average)	0.961	3.82	about 170*	13.7**	64.0
Hungary, Nagylengyel Well N° 267(1960)	0.983	5.24	170+*	16.9**	71.0

* Extrapolated value
** Conradson carbon residue

for asphalt manufacture. The production of identical quality asphalt is advisable when determining yield and quality of residues from a crude oil. A 100 penetration asphalt at 25°C is commonly made. Thorough laboratory methods(8,9) are known for the investigation of crude oils to be used in the first place for asphalt manufacture. Table 9 shows the research data of crudes of high asphalt content and % asphalt yield of 100 penetration at 25°C.(10)

2.3 Distillation

2.3.1 Crude Oil Distillation

The oldest and simplest processing of crude oil is by distillation. The distillation process consists of heating and evaporating some components, separating and condensing the vapors and final cooling of the distillates, a mixture of liquids -in this case of a crude oil(11). Crude oil distillation processes are commonly classified according to the method of evaporation. The following possibilities can be distinguished:

1. Batch evaporation where the material to be distilled is heated to gradually increasing temperatures and the components are evaporated in order of their boiling points. This working method is also called differential evaporation. Only a slight part of the vapors is in equilibrium with the liquid in this method,
2. Continuous evaporation is carried out, when the liquid mixture to be evaporated is heated to a temperature high enough to keep even the component with the highest boiling point in the gaseous

state. This method, also called equilibrium distillation differs from the previous method also in the fact that the vapors formed are not taken off from the liquid surface where they are formed but are left with the liquid during heating. A practical example is the heating in pipe stills.

The vapors evolved in the pipe still system are kept with the liquid and not distilled off, until they are expanded into the column. This evaporation method is much more advantageous than the first one, since the crude oil consisting of a great many components has to be heated to a considerably lower temperature to produce the identical amount of distillate.

Another advantage consists in avoiding cracking during evaporation of the heavier molecules in the crude oil, due to considerably lower distillation temperatures. Thermal cracking starts at 370°C.

Crude distillation processes may be divided according to the pressure also:

1. Atmospheric distillation, if the process is carried out at an atmospheric pressure. Most distillation processes belong to this group.
2. Distillations effected at reduced pressures are called vacuum distillations. The purpose is to reduce the distillation temperature by which cracking of high molecular weight material (lubricating oil distillates) is avoided.
3. Distillation under pressure is carried out at higher pressure than atmospheric. This method is applied for special requirements.

2.3.1.1 Atmospheric Distillation

As has already been mentioned, batchwise or fractional-, semicontinuous, and continuous distillation processes are distinguished in practice[12]. The batchwise or fractional distillation apparatus is

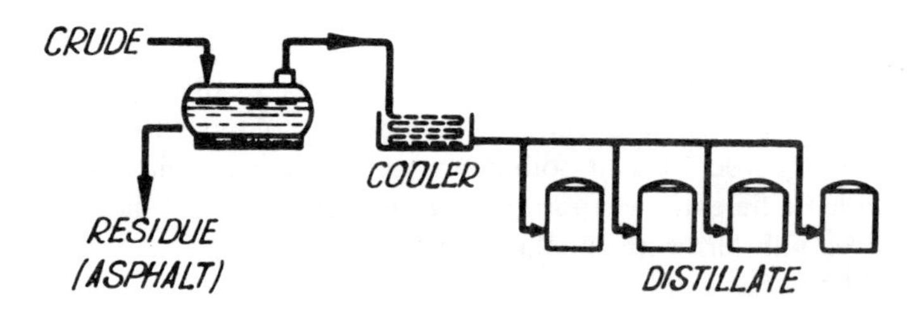

Fig. 10. Discontinuous shell still distillation

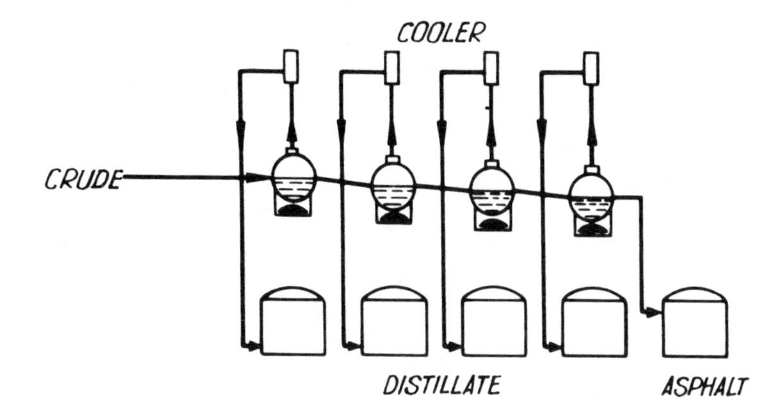

Fig. 11. Semi-continous battery distillation plant (Baku process)

characterized by the total quantity of the material to be distilled being fed into the shell still, and the vapors obtained by gradual heating taken off immediately and condensed continuously (Figure 10). Due to its various disadvantages, this process is seldom used in industry.

Compared to this process, the semi-continuous plant process or Baku process represents some progress. In this procedure a battery of 3-4 stills are connected in series. Crude is pumped in the highest still, and while the gasoline vaporizes, the rest flows into the next lower still by gravity. (Figure 11).

The distillates leaving the stills are collected separately. Although this process is more advantageous than the previous method, it is considered obsolete(13). At present the most commonly used system for petroleum distillation is the rectification tower (Figure 12).

Crude oils is charged through the heat exchanger in the pipe still and is heated at a temperature where all overhead fractions and the components to be taken off as side cuts vaporize. In practice, the material is heated to a somewhat higher temperature to ensure the reflux to the lowest trays too, since the separation of individual fractions is not sharp enough otherwise. The material heated in the still is expanded at the bottom of the fractionation column, where the vaporized part is separated from the liquid and moves upwards. Since a temperature decrease is involved in the expansion, the still outlet temperature is higher than that of the column. The maximum inlet temperature is

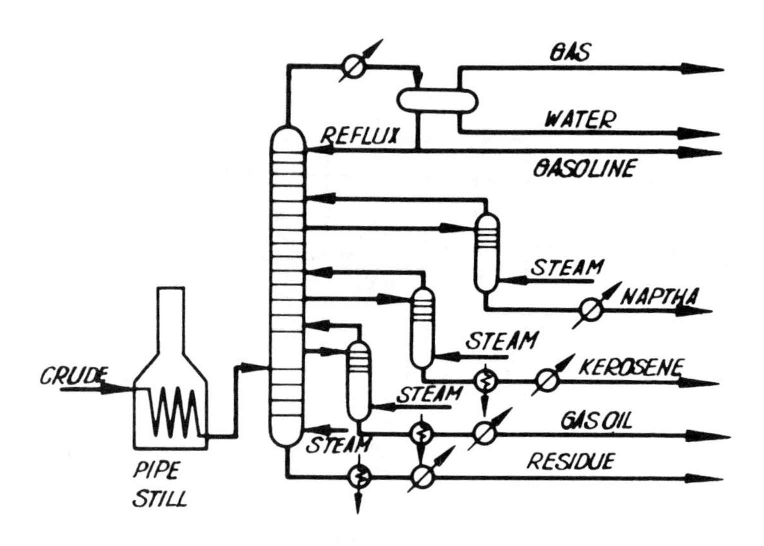

Fig. 12. One stage continous (pipe still) distillation unit

given by the initial temperature of cracking. Should the vaporization degree be increased without cracking, the partial pressure of oil vapors must be diminished by steam charged into the vaporization zone of the tower. Vapor is commonly blown into the residue at the tower bottom. The heavy distillates from the vapors moving upwards are condensed on the lower trays in the column, the lighter fraction on the lower trays in the column, the lighter fraction on the higher trays. Gasoil, kerosene, and naphtha distillates are collected on trays of suitable construction, from where they are passed into the stripping towers. The entrained undesirable light distillates are driven off by steam injection from the material flowing off the 4-5 trays mounted in these columns. The light distillates are then recycled into the fractionation column from the top of the stripper. The distillates taken off the strippers are first passed through a heat exchanger, and then through a water cooler. Finally they enter the receiver. The lightest fraction, gasoline, is passed from the column head into the condenser and cooled to normal temperature. The uniform removal of the residue accumulated in the tower bottoms is made by a level controller. The material is continuously exhausted by the residuum pump and passed into the storage tank through a heat exchanger and a water cooler.

2.3.1.2 Vacuum Distillation

It is of outstanding importance in the use of vacuum distillation in the petroleum industry that the boiling points of materials vaporized at reduced pressure are lowered. Thus high molecular weight substances which would undergo cracking under atmospheric conditions may be vaporized without thermal decomposition.

In the beginning vacuum units were operated batchwise, but modern units are run in a continuous manner[14]. Techniques and construction of vacuum units correspond approximately to those in continuous atmospheric units, as described above.

The decrease of distillation temperature brought about by the vacuum operation is enhanced also by steam injection. In industry, such units are operated most economically at a pressure of 30-50 Torr with steam injection. If no steam is injected and the plant is operated "dry", the absolute pressure has to be decreased to 1 Torr. Very expensive high vacuum equipment would then become necessary.

The most commonly used vacuum distillation procedure in practice is the one stage vacuum unit. The distillation plant consists of a feed pump, a pipe still, a vacuum tower, side stream towers, as well as of heat exchangers, coolers, product and recycle pumps. There are also the vacuum ejectors, and the necessary measurement and control devices. Such units are only applied for definite tasks such as the distillation of intermediary products in lubricating oil production and other special purposes.

The so-called two-stage units consisting of the combination of at-

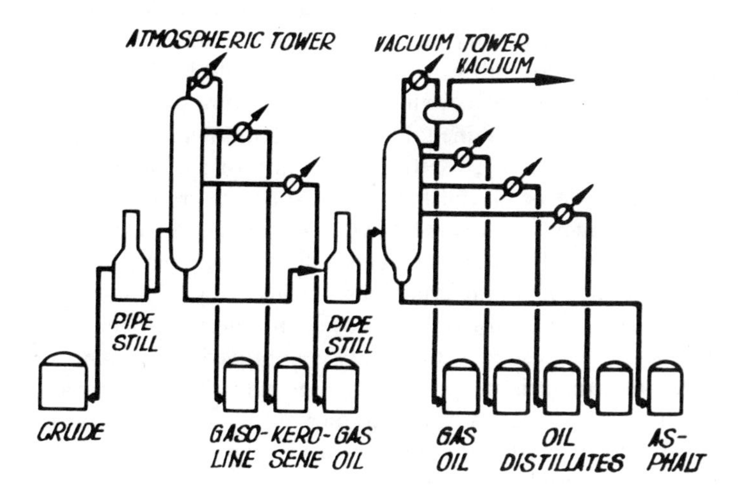

Fig. 13. Two stage distillation unit

mospheric and vacuum distillation are also commonly used (Figure 13).

In this unit, the residue collected at the bottom of the atmospheric tower is removed by a pump and charged through the pipe still into the vacuum tower upon heating at the suitable temperature. If the sulfur and salt contents in the crude are high, the so-called three-stage distillation is used to avoid corrosion damages. The first column is employed to remove hydrogen sulfide and water with some gasoline. A pressure of two atmospheres is maintained in the first column. The top product obtained is gasoline, water, and hydrogen sulfide. Recently a high vacuum tower is often used together with the atmospheric and vacuum unit to concentrate the vacuum tower residue.

2.3.2 Asphalt Manufacture by Distillation

The general distillation methods used in petroleum refining have been dealt with above. Different plant conditions must be chosen with crude oils of different origin and residues, according to composition, water, sulfur, salt, and metal contents, as well as the objective of processing operations, such as lubricating oils, crack products, asphalts. The primary aim is to obtain lubricating oil distillates of good qualities[15] in processing quite a number of crude oils.

At the same time asphalts are produced as a by-product in some cases, while in others only asphalt manufacture is aimed at.

Selection of processing methods and that of suitable plant conditions for these depend in the first place on the raw material composition. Its knowledge is necessary for plant design and later in operating the plant[16]. Thus above all raw material composition, the conditions of equilibrium distillation, as well as the temperature susceptibility of the crude oil must be determined. In sulfur containing raw materials, totai sulfur and the quantity, as well as stability of the individual sulfur compounds must be established. On heating the crude oil, the sulfur compounds decompose to evolve hydrogen sulfide. Corrosion problems are caused by the latter and effluents will be contaminated at the same time. With salt containing crude oils, the necessity of desalting should be considered, since not only corrosion problems are eliminated by salt removal, but also asphalt quality is improved. Salt quantity accumulated in asphalt is in some cases large enough to deteriorate electrical properties of asphalts, and to increase its water absorption. The pretreatment of crude oil comprises also the removal of naphthenic acids when the crude oil is neutralized with sodium carbonate, and the naphthenic soaps formed are precipitated in the residue. This has

TABLE 10.
Asphalt From Untreated Matzen Crude and From Crude Pretreated with Soda Solution

Sample *Soda treatment*	*A* —	*B* +	*C* +	*D* —
Penetration at 25°C, 0.1 mm	318	130	87	37
Softening Point, °C	34	45	52	53
Ductility at 15°C, cm	>100	—	—	—
at 25°C, cm	—	>100	>100	>100
Fraass breaking point, °C	−18	−18	−8	−7
Solubility in water, %	0.18	1.7	1.3	0.0
Acid value, mg KOH/g	0.66	10.28*	0.36*	0.32
Ash, %	0.11	0.86	0.83	0.14

* Alkali value

an adverse effect on asphalt quality. Comparison data on asphalts treated by this method and those of untreated (Matzen) asphalt are compiled in Table 10. It can be seen that the water soluble part of the neutralised asphalt exceeds the specifications.

Since the asphalt content of some crudes amounts to as much as 70%, it is natural that some plants concentrate on processing the crude oil only to obtain asphalt. With some crude oils of exceedingly naphthenic character, the crude oil can be reduced direct to asphalt. Asphalt of 100 penetration can be reduced from Guadalupe crude oil (California) at 343°C. Sometimes however, higher temperatures must be resorted to with crude oils of different character. Table 11 comprises the characteristic temperature data of atmospheric asphalt manufacture from Nagylengyel crude oil[19]. It can be seen that 400°C is required with this raw material to reduce to 100 penetration asphalt.

Plant temperatures in the vacuum distillation are given for the sake of comparison. They are about 100°C lower than that required

TABLE 11.
Characteristic Temperature Data of Atmospheric Asphalt Manufacture (Nagylengyel Crude Oil)

	Atmospheric			*Vacuum*
Distillation method	*1*	*2*	*3*	*4*
Pipe still outlet temperature, °C	365	392	400	280
Asphalt				
Penetration at 25°C, 0.1 mm	450	234	100	223
Softening point, °C	38	41	50	42

under atmospheric conditions. It should be mentioned that stripping steam was used in both cases. The high temperatures required in atmospheric distillation involve the danger of overheating and undesirable changes may occur. Thus if the application of this method is absolutely necessary, continuous strict control of the temperature and that of the produced asphalt quality is most important.

In general vacuum distillation is commonly used and necessary for the production of asphalts with various determined penetrations. Vacuum distillation of asphaltic residues does not differ to a great extent from the above described process, but it is somewhat simpler[20]. Fractionation in a vacuum tower need not give as sharp a separation as required for lubricating oil production, where difficulties arise in further treatment if the oil distillate fractionation was not sharp enough. The vacuum towers used mainly for asphalt manufacture possess therefore fewer plates. Corresponding to this, fewer side cuts must be taken off the tower simultaneously. The feed stock can be heated in such units to higher temperatures than those used in lubricating oil production without deterioration of asphalt quality.

The maximum process temperature is limited by the thermal decomposition or cracking. A maximum outlet temperature of 400°C is used in the vacuum pipe still at pressures from 30 to 80 Torr for asphalt manufacture in most units, and some of them are operated at even higher temperatures. The temperature to be used depends on the way of heating and on the feed stock properties. With a naphthenic feed stock at 40 Torr, a temperature of 340°C will be suitable to produce 100 penetration asphalt, whereas with paraffinic feed stock, 480°C is necessary to obtain the same asphalt quality under identical conditions. Unusually high vacuum must be applied in the last case to decrease the temperature.

Manufacture of the desired vacuum asphalt quality from various feed stocks is carried out by different distillation procedures in the individual plants. In a process similar to the atmospheric method, the feed stock preheated in the heat exchanger passes to the pipe still where it is heated to the desired temperature[22]. Then the heated material is introduced into an evaporator at a height of approximately two thirds of it. The asphalt flows off the stripping trays while superheated steam enters at the bottom. The asphalt maintained continuously at the same level is pumped off the evaporator bottom. The flashed vapors enter from the evaporator the fractionation column at the bottom of which superheated steam is charged. The fractionating column serves

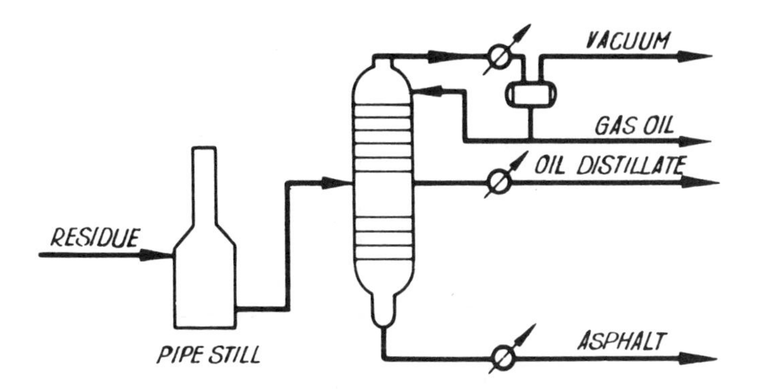

Fig. 14. Asphalt manufacture in a vacuum unit

to separate the flashed vapors into the desired fractions, such as gasoline, kerosine, and gas or fuel oil.

If asphalt manufacture is the principal object of the distillation, the vacuum flash system shown in Figure 14 will be used. At most two distillates are produced in this process apart from the residue. Its advantages are that the equipment is simpler and less expensive, the extent of the fractionation is smaller, and thus costs decrease. The flow sheet of a unit to process crude oil with 50% asphalt content is described by Kastens[23]. The crude oil is first charged from the tank into the heat exchanger system with the higher distillates of the fractionating tower and is then preheated to 280°C with hot asphalt taken from the tower bottoms. The heated crude oil is then passed into a flash drum, where water and lighter distillates are removed. It is here that the feed to the furnace is dehydrated so that the inorganic salts in the crude are precipitated and go through the heating coils in a solid state, thereby preventing the deposition of salt scale on the hot surface of the tubes. The water and light ends driven off and leaving the drum as overhead, the distillate is passed directly into the fractionation tower, while the dehydrated crude oil is heated to 280°C by a further heat exchanger and then charged into the pipe still. The crude oil is heated to 360°C in the still and expanded into the fractionating tower with 21 trays. Two side cuts are taken off the vacuum tower apart from the asphalt bottoms and the overhead naphtha or gasoline fraction. Stripping steam is introduced into the tower bottom. The flow sheet of this distillation method is shown in Figure 15.

As it has been mentioned above, asphalt is obtained among other

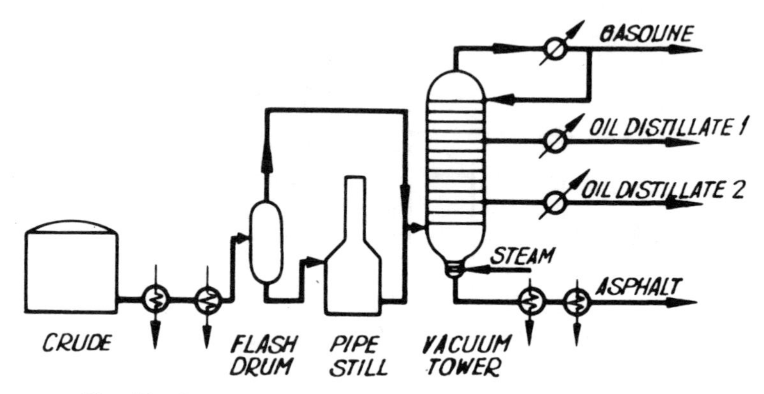

Fig. 15. One stage vacuum distillation unit with flash drum

products both in atmospheric and vacuum towers. These units differ depending on whether the principal product is lubricating oil and asphalt as a byproduct only, or whether asphalt manufacture is the chief object of the operation[24].

Also vacuum residue asphalts with a softening point of 80°C are made by the processes described above from the atmospheric bottoms, which in turn depend on the properties and heat treatment of the crude oil. Various processes are known to obtain straight asphalt with higher softening point. A distillation tower operated by a circulation

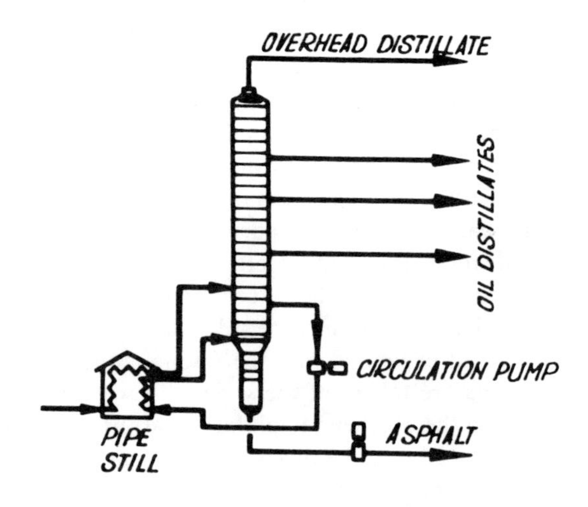

Fig. 16. Asphalt manufacture in circulation system

system to produce high softening point asphalt can be seen in Figure 16. The low softening point product from the tower bottom is circulated through the pipe still by a pump. Steam is injected into the preheating coils to avoid coking. A further quantity of the heavy oil constituents vaporizes in the lower third part of the tower from the asphaltic residue[25].

The separation of asphaltic residues from the heavy oils is made by so-called carrier-distillation process.[26] Should good quality lubricating oil result, its quality would be deteriorated even by slight cracking. If the feed is blended with a light distillate "carrier" in these cases, e.g. with gas oil, the temperature of equilibrium vaporization will decrease. The reason for this is that the partial pressure of the heavy oil parts diminishes in the presence of gas oil vapors and more heavy oil is vaporized than without gas oil at a given temperature. In plant practice, 100 to 200% gas oil is mixed with the distillation residuum of the crude oil and the mixture is evaporated in vacuum. The gas oil "carrier" is separated overhead, while a very heavy oil distillate is obtained as side cut. High penetration asphalt with high softening point is produced in the process as bottoms. In special cases asphalts are manufactured in high vacuum units (from 1 to 5 Torr)[27].

In the process shown in Figure 17, the residue is passed to a preflash tower and then distilled in dry vacuum. Flashed gas oil is condensed inside the vacuum vessel by direct contact with circulating oil. Pressure drop between the jet suction and flash zone is held unusually low by careful attention to internal design. Vacuum jets used in the high vacuum systems are of low volumetric capacity. In effect this is a built-

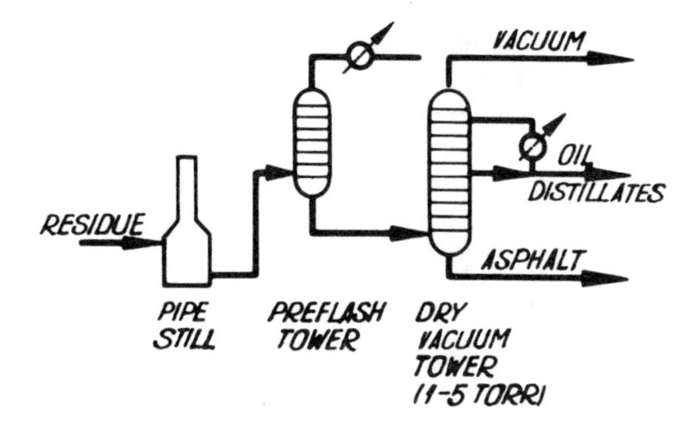

Fig. 17. Distillation unit with high vacuum tower

in way to regulate temperature. Should cracking occur, the jets overload and corrective measures are taken.

Care must be taken when operating vacuum units to maintain the pressure low and constant at any rate, since any pressure fluctations cause product quality deterioration, decrease in yields, and influence the whole process adversely. The apparatus and the flanges of the coil system must be completely airtight, since air penetration from outside will deteriorate the vacuum. These units must therefore be maintained with greater care than atmospheric units.

2.3.3 Properties of Straight Run Asphalts

The larger part of asphalt is manufactured by distillation according to the above described methods. Apart from the commonly used name of straight run or straight reduced asphalts, various terms are applied to distillation asphalt with a softening point higher than 80°C. These are called high vacuum asphalt due to the high vacuum required for their production. However, it must be mentioned that apart from this process widely used in certain countries, asphalts of softening points above 80°C can be produced from high asphalt content crude oils without high vacuum by adequate changes in distillation conditions, chiefly by increase of temperature.

Asphalts of variable penetration and softening point are obtained from the vacuum residue of crudes intended in the first place for lubricating oil production.

Table 12 illustrates this fact since it comprises data for asphalts made from the industrial lubricating oil unit residues at 40 mm Hg. The re-

TABLE 12.
Straight Run Vacuum Residue Asphalts Available From Lubricating Oil Production

	Asphalt	
Crude	*Softening point °C*	*Penetration at 25°C 0.1 mm*
Anastasievka (Soviet Union)	38–46	270–98
Lispe (Hungary)	52–63	109–37
Matzen (Austria)	42–53	204–60
Romashkino (Soviet Union)	44–55	246–30
Tuimaza (Soviet Union)	43–54	230–33
Tyulenovo (Bulgaria)	non measurable viscosity 35–40°E at 100°C	

TABLE 13.
Correlation Between Vacuum Distillation Temperatures and Characteristic Data of the Asphalt (Nagylengyel Crude)

Sample	*1*	*2*	*3*	*4*	*5*
Pipe still outlet temperature, °C	280	320	330	340	388
Asphalt					
Penetration at 25°C 0.1 mm	223	74	33	15	12
Softening point, °C	42	53	65	76	85

sidue sometimes meets the specifications for paving asphalts and can be marketed. If not it will have to be further treated by blending or blowing. When the quality of straight asphalt reduced from this crude oil is according to standard and the manufacture economical or when the reduction of the crude oil seems to be necessary from the national economic viewpoint, a special distillation unit to produce asphalt will be erected in addition to the units required for lubricating oil production. Either atmospheric residue is processed in this unit, or the vacuum residue is reduced under different conditions to various penetration grade asphalt[(28)]. Process techniques are simpler with the high asphalt content crude oils used in the first place for asphalt manufacture. No obstacle is encountered to make asphalts of lower penetration and higher softening points[(29)].

The most important data of asphalt reduced from Nagylengyel crude oil at 40 Torr and various pipe still outlet temperatures are compiled

TABLE 14.
Characteric Data of Asphalt Reduced by Atmospheric Distillation (Nagylengyel Crude)

Sample	*1*	*2*	*3*	*4*
Distillation method	*atm.*	*vacuum*	*atm.*	*vacuum*
Penetration at 25°C, 0.1 mm	165	162	100	76
Softening point, °C	44	44	51	51
Ductility at 25°C, cm	100	100	100	100
Fraass breaking, point, °C	−20	−24	−17	−18
Loss on heating, %	0.2	0.2	0.4	0
Softening point after loss on heating, °C	50	48	54	56
Penetration at 25°C after loss on heating 0.1 mm	100	108	77	59

in Table 13[30,31]. It is obvious from the data that asphalts having lower penetrations and higher softening point result by temperature increase.

It has been mentioned that asphalts can be made from high asphalt content crude oils also by atmospheric distillation. Characteristic data of asphalt reduced under atmospheric conditions from Nagylengyel crude oil are shown in Table 14. For the sake of comparison, data for asphalt of similar penetration and softening point made at 40 Torr are also given here[32]. It can be seen that the standard values of outstanding importance are practically identical for both atmospheric and vacuum asphalts. Determinations of paraffin wax contents which were not included in the Table have also similar values for the asphalt made by both methods.

To illustrate the changes in properties of the asphalt produced from crude oil by vacuum distillation, data for Nagylengyel straight run vacuum asphalt are compiled in Table 15[33]. It is obvious from the Table that by penetration decrease and softening point increase the ductility, the Fraass breaking point, and paraffin wax content of the asphalt diminish, whereas its specific gravity, viscosity, sulfur and asphaltene contents increase. Apart from the changes in standard data, the change in asphalt group composition is of interest. Changes in res-

TABLE 15.
Changes in Asphalt Properties by Vacuum Distillation (Nagylengyel Feed Stock)

Sample	*1*	*2*	*3*	*4*	*5*	*6*
Penetration at 25°C, 0.1 mm	286	162	76	30	17	9
Softening point, °C	39	44	51	65	74	88
Ductility at 25°C, cm	100	100	100	56	8	0
Fraass breaking point, °C	−28	−24	−18	−8	0	+9
Specific gravity at 25°C, g/ml	—	1.026	1.038	1.044	1.050	—
Viscosity °E						
at 150 °C	14.5	—	34.8	—	—	—
at 175 °C	7.3	—	15.8	37.0	—	—
at 185 °C	5.7	—	11.3	—	—	—
at 195 °C	—	—	—	17.6	33.7	—
Paraffin wax, %	2.2	1.9	1.5	—	1.2	—
Sulfur, %	—	5.0	—	5.1	5.7	—
Asphaltenes (n-Heptan), %	—	23.8	—	—	30.9	—

* at 15°C

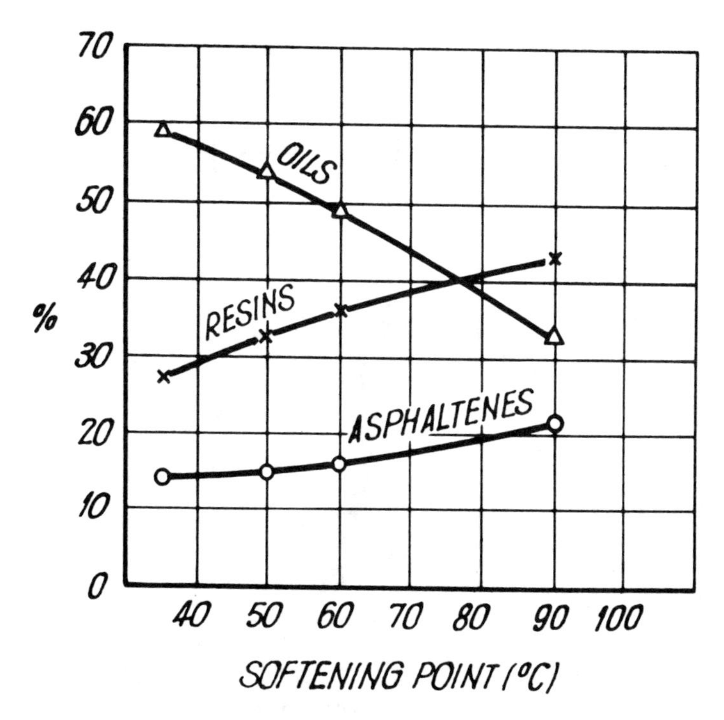

Fig. 18. Changes in asphalt group composition during distillation (Nagylengyel feed stock)

inous, oily parts, and asphaltenes of Nagylengyel asphalt are shown in Figure 18. The increase in asphaltene and resinous components and the decrease in oily components as the distillation progresses are emphasized in the investigations of Nyul, Zakar, and Mózes[34]. The processed crude oils and the straight run asphalts are characterized by the correlation softening point-penetration. Data referring to this in several known asphalts are compiled in Figure 19[35]. Data for straight run asphalts of different origin and of similar penetrations are given in Table 16. It is illustrated by this compilation that according to their different origin, the softening-point-penetration data of the individual asphalts do not corelate, nor do the other characteristics[36].

As it has been mentioned, the 100 penetration value at 25°C (100 g, 5 sec) is commonly used to compare asphalts made by distillation. Important test data are shown in Table 17 based on the work of Stanfield and Hubbard cited above[8]. The crude oils of different

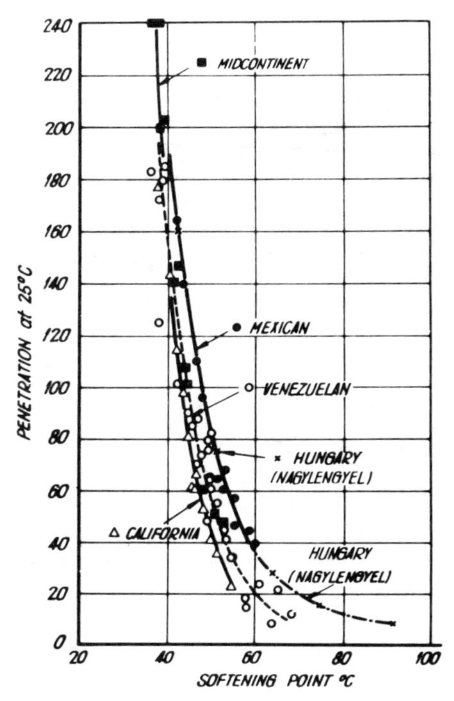

Fig. 19. Correlation between softening point and penetration of asphalts of various origin

TABLE 16.
Comparison of Straight Run Asphalts of Various Origin

Oriin	*Matzen*	*Romash-kino*	*Nagylen-gyel*	*Matzen*	*Tuimaza*	*Nagylen-gyel*
Penetration at 25°C, 0.1 mm	130	153	160	37	32	53
Softening point, °C	45	46	44	53	54	56
Ductility at 25°C, cm	>100	>100	>100	>100	>100	>100
Fraass breaking point, °C	−18	−10	−24	−7	−3	−13
Paraffin wax, %	1.1	2.0	2.0	0.9	2.0	1.5
Sulfur, %	0.4	3.4	5.0	0.44	3.0	5.0
Asphaltenes (Hexan), %	5	7.5	21	6	9	23

TABLE 17
Characteristic Data of American Asphalts with the Identical Penetrations but Different Origin

	Wyoming Elk Basin	*Montana*	*Wyoming Oregon Basin*	*Arkansas Irma*	*California Pondera*	*Mexico Tampico*
Asphalt yield, weight %	15.9	21.2	44.6	49.7	54.6	73.6
Composition:						
Asphaltenes %	14.5	20.0	20.0	22.5	12.5	29.5
Oils %	53.0	50.0	47.0	47.5	47.5	43.5
Resins %	32.5	30.0	33.0	30.0	40.0	27.0
Carbon residue, %	25.3	24.0	20.1	22.9	14.2	21.7
Sulfur, %	3.77	4.58	4.81	3.50	1.23	6.12
Specific gravity at 25 °C	1.024	1.031	1.026	1.018	1.010	1.036
Penetration at 25°C	100	100	100	100	100	100
Softening point °C	43.9	44.4	45.0	45.0	42.2	47.2
Ductility, cm at 25°C	100+	100+	100+	100+	100+	100
Viscosity at 98.9°C						
Furol sec	920	1250	1300	1900	730	2700
Kinematic stokes	19.9	27.0	28.1	41.0	15.8	58.3
Absolute poises	19.4	26.5	27.4	39.8	15.2	57.2
Viscosity at 135°C						
Furol sec	120	160	160	210	83.0	290
Kinematic stokes	2.95	3.45	3.45	4.53	1.79	6.26
Absolute poises	2.47	3.32	3.30	4.30	1.68	6.04

origin are distilled in the laboratory equipment under similar conditions provided for ensuring comparative evaluation. The characteristics of 100 penetration asphalts were estimated from the data of various penetration asphalts obtained from the individual crude oils. Also the yields of asphalts referred to the crude oils are given in the Table. The differences between the distilled asphalts of similar penetration but of different origin are illustrated well by these data.

Marketed straight run asphalts are classified by their penetration ranges. The classification used commonly will be taken up in the part dealing with paving asphalts. To characterize the straight run asphalts having high softening points, usually above 80°C, called also high vacuum asphalts, softening points limits are given[37]. The characteristic data of such asphalts as commonly marketed in West Germany are compiled in Table 18. As can be seen from the previous Tables and from Figure 19, all asphalts with higher softening points exhibit lower penetrations. The penetration value corresponding

TABLE 18.
High Vacuum Asphalts

Designation	*85/95*	*95/105*	*130/140*
Softening point, °C	85–95	95–105	130–140
Penetration at 25°C 0.1 mm	5–11	4–7	1–3
Ductility at 25°C, cm	2–5	0–20	—
Ash, max. %	0.5	0.5	0.5
Paraffin wax. max. %	2.0	2.0	2.0
Flash Point, min. °C	330	330	330
Specific gravity at 25°C, g/ml	1.04–1.07	1.05–1.08	1.06–1.1

to a given softening point is a function of crude oil origin. Considering that the paving asphalt standards of each country state the softening point together with the penetration range, it may be necessary in some cases to set the required softening point-penetration values. This is made possible by the blending of the processed crude and the atmospheric residue with another product in definite proportions, and by distillation.

It was proposed by Berkes, Ney and Luter[38] for gel asphalt manufacture to distill the crude oil designed for sol asphalt recovery together with a crude oil of suitable quality. It is possible to increase considerably the penetration of an asphalt with higher softening point, depending on crude oil and blended feed stock. Thus an asphalt with a softening point of 85°C and from 40 to 65 penetration could be made. Distillations of crude oils of various origin, with different but high asphalt contents under various blending proportions were investigated by

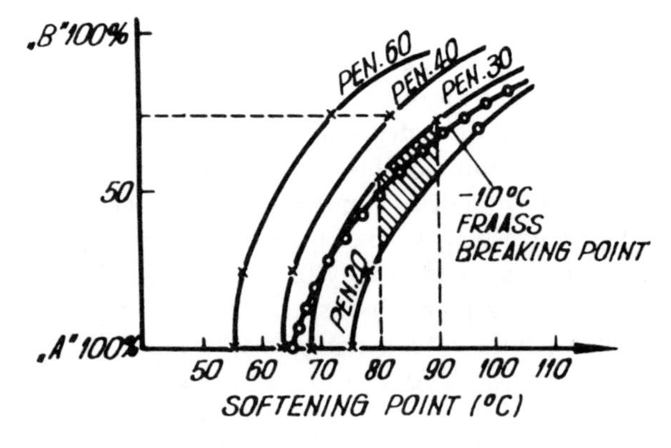

Fig. 20. Correlations between crude composition and asphalt quality

Zakar, Mózes and co-workers[39]. Based on these tests, the blending proportion range could be determined to obtain asphalt with softening points from 80 to 90°C and having 20-30 penetrations. Straight run asphalts were thus obtained corresponding to 82/25 blown asphalts. These correlations are shown in Figure 20. Besides percentage composition of the crude oils and relationship between softening point-penetration, the Fraass breaking point of the above mentioned asphalt is also included.

2.4 Extraction

2.4.1 Propane Deasphalting

The manufacture by distillation of high viscosity oils such as aviation and cylinder oils requires such high temperatures as to cause thermal cracking of the oil to be processed.

Extraction processes must be resorted to for the production of similar oils. Investigations for suitable solvents showed that various liquid hydrocarbons, such as propane, dissolve oil and paraffin wax to a great extent, but are very poor solvents for the asphaltic constituents. Thus a variety of separation processes by selective solvents was developed for lubricating oil production in which the oily components are separated from the asphaltic components due to the good solubility of the former. The method is called solvent deasphalting and is mostly applied in industry as propane extraction i.e. propane deasphalting[1,2,3].

Approximately from 30 to 70% oil and from 70 to 30% asphalt may be obtained from the feed stock by this method, depending on feed composition and operation conditions. Although this process was not developed for asphalt manufacture, it has become quite important for it.

The properties of the propane asphalts depend not only on the nature of the feed stock, but also on the type of the solvent used and on plant parameters such as temperature, pressure, oil: solvent ratio. A relatively large quantity of very soft bituminous material is precipitated by ethane from a given oil residue. If butane is applied instead of ethane, an asphalt with higher softening point will be gained in smaller yields. The effect of propane is between that of these two solvents. Interchanging solvents is not common in industrial processes, and propane is generally used, since this is the best solvent both from economical and from plant operation viewpoints. Considering the feed stock, the most favourable refining conditions are selected by adjusting oil-solvent ratios[4].

Industrial propane deasphalting was originally carried out as a batch onestage operation. The material to be treated was mixed with liquid propane at a fixed temperature in a closed and pressurized system, and then allowed to settle. Upon removal of the precipitated asphalt, the solvent was recovered. Although the bulk of the asphaltic component could be isolated from the feed stock by this process, no oil of desired quality could be obtained. In the course of improving this method, the asphalt separated in the previous process is washed with fresh propane in countercurrent flow and the solvent is then used to de-asphalt the next oil portion. Instead of batch multistage de-asphalting with propane, continuous countercurrent columns are used in large scale operations, similar to the other solvent refining processes. Mixing is thorough between propane and oil flowing countercurrent in these columns, and the components are separated from one another continuously[5,6]. In a modern plant, the feed stock to be treated is charged from the feed tank by a pump into the upper part of the extraction column passing a steam preheater. On the other hand, liquid propane is pumped into the bottom of this column. The oil is extracted by the ascending propane while the asphaltic phase precipitated out of the feed descends to the bottom where it is concentrated. The deasphalted oil solution containing propane is passed to a propane vaporizer heated with low pressure steam where the bulk of propane is removed. The separated propane flows through a cooler back into the propane tank. Most of the remaining propane is removed from the oil still containing propane in a vaporizer, heated by high pressure steam. To remove the last traces of propane, the oil is passed through a stripper where it is freed entirely from propane and is charged by pump into the tank for

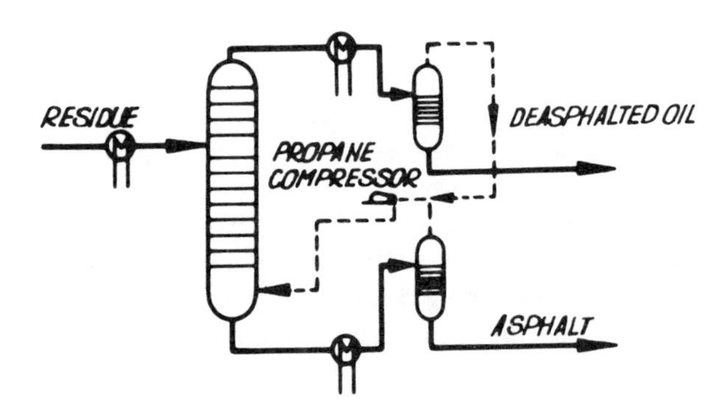

Fig. 21. Flow sheet of a propane deasphalting unit

TABLE 19.
Straight Run and Propane Asphalts With Identical Softening Points (Nagylengyel)

Method of processing	*Vacuum distillation*	*Propane de-asphalting*
Softening point, °C	72	74
Penetration at 25°C 0.1 mm	19	9
Ductility at 25°C, cm	9	0
Fraass breaking point, °C	0	13
Specific gravity at 25°C, g/ml	1.053	1.079
Paraffin wax, %	1.6	0.8

deasphalted oil. The asphaltic phase bottoms are drawn from the column, heated in the pipe still and led to the evaporator. Asphalt from the evaporator bottom is charged to the stripper where the removal of the last traces of propane takes place. Propane removed from the evaporator is collected in a suitable system and recycled into the propane tank by means of a compressor passing a cooler. The flow sheet of a propane deasphalting unit is shown in Figure 21.

The feed stock treated in the deasphalting unit is a vacuum residue, in many cases corresponding to an asphalt with higher penetration. The oily constituents are then separated from the soft asphalt and a lower penetration asphalt having smaller oil content is produced. This is proved by the comparison data in Table 19. Quality differences of straight run and of propane asphalts from Nagylengyel having the same softening points show the effect of propane deasphalting on asphalt properties[7]. Characteristic data of various propane asphalts from Romashkino feed stocks investigated partly in pilot plant and

TABLE 20.
Propane Asphalt from Romashkino Crude

Method of processing	*Vacuum distillation*	*Propane asphalt*				
		1	*2*	*3*	*4*	*5*
Softening point, °C	44	53	57	69	71	75
Penetration at 25°C 0.1 mm	246	42	26	8	7	5
Ductility at 25°C, cm	100	100	100	11	0	—
Fraass breaking point, °C	−15	−5	0	+7	—	—
Paraffin wax, %	2.0	—	2.0	—	—	—
Sulfur, %	3.3	3.59	3.61	—	3.59	—
Asphaltenes (Hexan), %	5.0	6.5	10	—	15	—

partly in the laboratory are shown in Table 20, based on the work of Zakar and Mózes[8]. The feed stock for propane extraction unit is given in the first column of the Table, out of which all the other products were obtained under various conditions. A wide range of products can be obtained during this process from paving asphalt[9,10] to brittle qualities which are easily ground depending on the feed stock treated and on production conditions[11]. The quality of the produced asphalt is however a function of the conditions set for lubricating oil manufacture. At best a good and easily marketable asphalt is produced along with the desired lubricating oil. Should the bituminous material not meet the requirements, further processing would become necessary. Then blending with other components is resorted to (12).

The asphalt made by the Duo-Sol method belongs to the extract asphalt group. This process is also applied to produce lubricating oil. Beside propane used to precipitate asphalt, the resinous components which are undesirable for lubrication are also removed by a phenol-cresol mixture as solvent. The resinous parts are concentrated in the asphalt[12].

2.4.2 Extracts from Solvent Refining of Lubricating Oils

The term "solvent extract asphalt" means the asphalt produced from the extract obtained in the solvent refining of lubricating oils as with furfural and phenol. The material is used above all as a blending component. Its properties depend to a great extent on the applied feed stock, the solvent, and the raffination grade[14,15]. Kossowicz[16] describes the application of asphalts obtained from furfural or phenol extracts of paraffinic and naphthenic materials. Blowing the liquid and semi-liquid lubricating oil residues at 250-300°C, asphalts of various softening points can be made. Good quality solvent extract asphalts

TABLE 21.
Asphalts Manufactured by Distillation and Blowing From the Extract of Lubricating Oil Solvent Refining

	Distillation				
	A	*B*	*Blown products*		
Specific gravity at 25°C, g/ml	1.030	—	—	—	1.046
Softening point, °C	29	55	42	58	69
Penetration at 25°C, 0.1 mm	—	22	80	17	9
Ductility at 25°C, cm	—	100	100	56	5
Fraass breaking point, °C	—	+10	−9	−2	+3

TABLE 22.
Blown Asphalts from the Romashkino Vacuum Residue of Lubricating Oil Extracts

	Vacuum residue of extract	*Blown products*			
Softening point, °C	34	58	77	84	109
Penetration at 25°C, 0.1 mm	—	20	5	4	1
Ductility at 25°C, cm	—	100	0	0	0
Fraass breaking point, °C	—	+1	+13	+18	+30
Paraffin wax, %	—	1.2	—	—	1.0

to be used blended or without blending are obtained by this method from products with low softening points up to asphalts with softening point of 160°C.

Zakar showed[17] that solvent extract asphalt could be manufactured from lubricating oil refining extracts by vacuum distillation. Depending on the distillation grade, softer or harder asphaltic residues result. The soft residue can be blown at will. Characteristic data of solvent extract asphalts are shown in Table 21. These were made in plant experiments from lubricating oil extract blends from Matzen and Lispe.

The vacuum residue of lubricating oil extracts blown as proposed by Zakar (18) resulted in asphalts, the characteristic data of which are shown on Table 22. Solvent extract asphalts can be used for special requirements or as suitable asphalt blending components due to their group compositions which differ to a great extent from those of conventional asphalts.

2.5 Blowing

Only slight quantities of asphaltic components are contained in some crude oils. Asphalt is generally produced by blowing the raw material with air in such cases. Again, so little high boiling gas oil can be obtained from other crude oils, that the vacuum distillation commonly used will not prove economical. These crude oils are often treated by air-blowing, by means of which, for example, quality paving asphalts are obtained. If a special asphalt is required, air-blowing will be used.

The asphaltic components of the processed crude oil remain practically unchanged during distillation and solvent extraction. On the contrary, under certain conditions alterations are caused in the treated raw material and the asphalts by the influence of air. when blowing is used. Most important changes of very complex character take place in the partial oxydation of hydrocarbons[1,2]. Formerly blown asphalts were called oxidized asphalts. There is some reason for this term, but it is still a misnomer, since the term oxidised is used in practice to denote that the product consumes considerable oxygen quantities, which does not happen in this case.

2.5.1 Fundamentals of Asphalt Blowing

Blowing of asphalt is a heterogeneous reaction between a gaseous inner phase (air, air $+O_2$ or O_3; air $+Cl_2$, air $+NO_2$) and a liquid outer phase (petroleum hydrocarbons). To accomplish this process, the necessary reaction conditions are to be brought about in a suitable reaction vessel (reactor). A vertical cylindrical vessel is selected in most cases such as a circulation vessel or a flow tube.

The blowing process may be understood by investigating the reaction conditions, the chemistry, and the kinetics of the reaction.

2.5.1.1 Reaction Conditions

Material transfer brought about by blowers. With two small air feeds, the formation of bubbles is also slow and it depends only on gravitational forces, static pressures, and the surface tension. Bubble size is given by the blower diameter[3]. With increasing air feed (about 9 m/sec air rate at the blower) so called quick blower gas is achieved. Since a determined constant maximum bubble frequency belongs to each blower, being the smaller, the larger the blower diameter, the bubble volume is increased, by increasing the air supply. Moreover, the bubble size becomes a function of viscosity. Further increase of air feed results in jet gases[4]. To provide for a large enough material transfer exchange surface, a great number of blowers are required which are fitted in perforated plates, perforated rings, spiders, or sinter plates.

In producing the material transfer surface by outside mixing, an air jet entering the liquid is fed towards a baffle plate through a blower or several blowers with large diameters, and the resulting air bubbles of various size are finely dispersed by mechanical mixing. The average bubble diameter to be achieved, which determines the material exchange surface is independent of the viscosity within a wide range and is determined only by specific gravity, surface tension and turbulence (power absorbed per unit volume). Since the power absorption of the mixing device decreases strongly with increasing air feed at determined revolution number[5], an optimum run may be obtained by the utilisation of a torque modifier. The air bubbles produced by the blower (perforated plate, mixing device etc.) result in a dispersion with the liquid under holdup formation. The individual small bubbles ascend to the dispersion surface and cause an intensive circulation of the liquid. The dispersion is constant only if the distance between two small bubbles equals at least 2 d, or bubble coalescence occurs resulting in a smaller exchange surface, and the viscosity of the liquid does not exceed 70 cP (6). This limiting stability condition $s=2d$, is not depending only on the gas content of the dispersion. The bubbles are pushed towards the middle by the liquid flowing backwards from the vessel wall, the bubble shape (ellipsoid) makes the bubbles capsize, and they wander towards the middle of the vessel. Outrunning phenomena are due to various ascending rates. After a determined ascending time (ascension height) only coarse bubbles prevail.

In practice, very interesting consequences may be drawn as to the vertical cylindrical vessels from the above discussed circumstances. The blower diameters and blower distances have to be adjusted to the feed stock and to the desired air feed (average viscosity at blowing temperature). The effective vessel diameter is not constant and it is a function of gas diffusion. It is useless to exceed a determined optimum conditions of vessel height, and so are excessive air feed, sinter plates or candle filter, countercurrent processes, or circulation pumps. In circulation pumps, with separated zones for ascending dispersion and descending liquid the dead space is eliminated at the interfaces. The height of a vessel with outside drive at any desired height may be the multiple of the optimum height in a vessel operated by blowers, due to the possible redispersion.

Two possibilities are available to influence reaction times— decrease of individual small bubbles (increase of ascension time) by pressure and finer dispersion or lengthening of bubble paths. Both measures are applicable only within very narrow ranges, as mentioned before. To produce the desired temperature does not present any difficulties, nor does its maintenance, since cooling becomes necessary due to the heat of reaction. Pressure maintenance is only a question of apparatus.

2.5.1.2 Chemistry and Physics of the Asphalt Blowing Process

Our present knowledge on blowing may be summarized as follows: The asphalt hydrocarbons contact the oxygen of air to form water with part of the hydrogen. The water is discharged from the system as steam. The hydrogen content of the asphalt residue is diminished and it becomes, therefore, more unsaturated than the feed. The unsaturated molecules react chemically and reduce the hydrogen content of the blown material during this process, and a slight quantity of oxygen is linked. Blowing of asphalt is an exothermal process. The heat evolved during it is a function of the chemical nature of the treated feed, of the degree of oxidation, and of the temperature at which blowing is commenced.

Goppel and Knoterus(7) confirm the fact observed by other investigators, that dehydrogenation also takes place in blowing apart from the reaction of asphalt with oxygen. Oxygen combines with asphalt to form esters, hydroxides, carbonyl compounds, and acids. More than 50% of the oxygen enters the ester group, and the remaining part is divided uniformly among the other groups.

The quantity of oxygen chemically bonded decreases in the blown asphalt with increase of blowing temperature, and increases with increasing aromaticity of the feed. The C-C linkage is formed at the optimum temperature of 250°C. Ester formation is considerable at lower temperatures, requiring more oxygen. At higher temperatures reactions prevail which do not favor formation of linkages.

The following four reaction types were detected:

1) Reaction decreasing the molecular size, producing blowing distillates, H_2O, and CO_2,
2) Reactions where molecular size changes are negligible; H_2O results,
3) Reactions increasing molecular size to produce H_2O, CO_2, and asphaltenes,
4) Concentration processes in which asphaltenes are concentrated as blowing distillates are driven off.

The consecutive character of the asphalt blowing process may be proved by observation of the reaction heat of the overall blowing process. Primary products seem to be the peroxides and/or hydrogen-peroxides shown by infrared spectroscopy(8). According to Goppel, mainly CH-groups arise in the initial period (about 40 cal/mole). The temperature increase/time ratio is small. During the main reaction period when acid- and ester formation has been well established with about 80 cal/mole, the temperature/time ratio is constant and decreases only when the quantity of reactive material becomes smaller or coalescence takes place caused by exceeding the 70 cP limit. (With Austrian blowing feed from Matzen this is experienced at Ring and Ball softening point of approximately 120°C, and a n-heptane asphaltene content of 32% (Senolt, private communication.)

To study the changes, the asphalts in question were divided into asphaltenes, hard and soft resins, and oily constituents by Hoiberg and Garris[9]. It was stated that upon continuation of the blowing process the asphaltene content increases, the quantities of resins decrease. Oily constituent contents are changed only very slightly.

It is confirmed by the experimental data of Sergienko, Delone, Krasavtchenko, and Rutman[10], that with the progress of blowing, the quantity of resinous and asphaltene containing materials exhibit a parallel increase, whereas oily constituent quantities diminish at the same time. During blowing the resin content remains practically constant, the oily hydrocarbon constituent decreases continuously, and the asphaltene content increases to a similar extent. This indicates that resin formation from oily hydrocarbon materials represents a transitory process towards the change into asphaltenes. When the ashaltene concentration of the blown asphalt reaches 35-40%, the carbene and carboid formation becomes considerable.

The products from naphthenic asphalts formed in the blowing process were studied by Moiseikov and Starobinec[11] as a function of time. The changes of the important characteristics (penetration, ductility) may be divided into three phases. Strong diminution of penetration and ductility takes place in the first phase, the decrease of these values becomes gradually smaller in the second phase, and they are stabilized in the third phase. These changes are in agreement with the investigation results on group composition of blown asphalts.

Blowing effects on Matzen, Lispe, and Nagylengyel asphalts were studied by Nyul, Zakar, and Mózes[12]. Two gasolines with different aniline points were studied in the separation. Great differences between Nagylengyel asphalt on the one hand, and Lispe and Matzen asphalts on the other hand were experienced during the blowing process. With Nagylengyel asphalt, the same changes took place when these two solvents were used. Beside enrichment of the asphaltene constituent, resinous and oily parts diminished. Again, with Lispe and Matzen asphalts with small asphaltene contents, transitory increase of resinous components was found in the first blowing period, when the gasoline with the smaller aniline point was used. Asphaltene formation was also less in this initial period. This difference could not be experienced when the solvent having the higher aniline point was used.

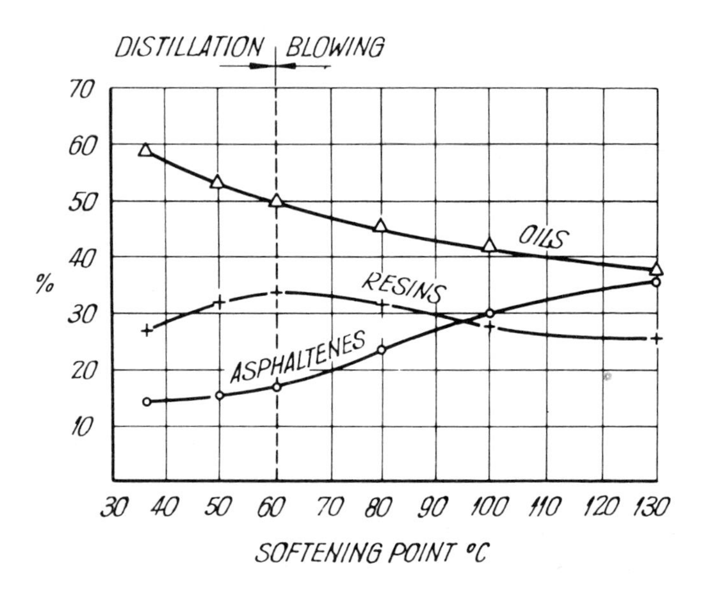

Fig. 22. Group composition changes of Nagylengyel asphalt during distillation and blowing

The changes in group composition of Nagylengyel asphalt during blowing was shown in Figure 22. For the sake of completeness, the group composition changes in the straight run asphalt blowing feed during the previous distillation are also given.

When blowing materials of the same origin but of different consistencies, the group composition of the blown asphalts change similarly and are almost parallel. Group composition changes of Nagylengyel asphalts having different softening points are compiled in Figure 23.

Blowing solvent extracts were also investigated by Kossowic[13] in his study on extract asphalt properties. As regards group compositions, these materials are remarkable by their high oil contents, Beside diminution of oil content and growing asphaltene content commonly experienced, attention is drawn to the different behavior of resinous constituents. Comparative tests based on the method of Lyshina for group composition study indicated, that during blowing asphalts of various origin, also the quantity of resinous constituents decreases. Again, when blowing extracts with small asphaltene and resin contents, the resinous component quantity increases as a function of the feed and then remains practically constant.

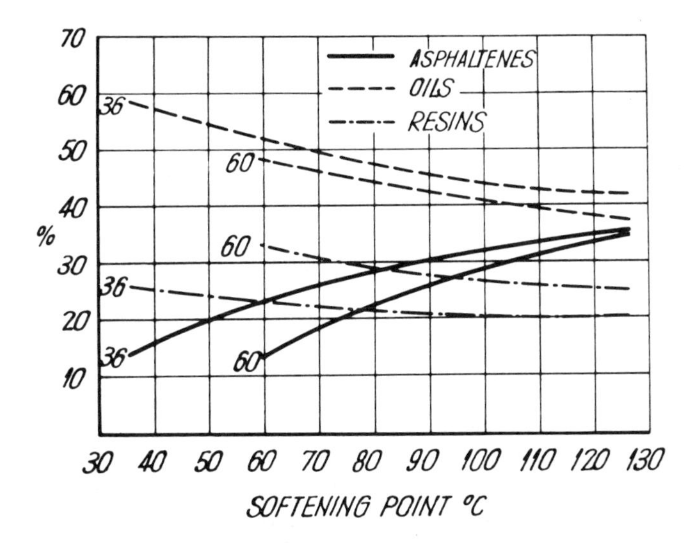

Fig. 23. Group composition changes of feed stock of the same origin (Nagylengyel) and various consistency. The figures on the curves indicate the softening point of the feed stock in °C

It was stated by Pentchev[14] studying the blowing at various temperatures of 30% Tjulenovo residue, that the blowing effect at 250°C is considerably smaller in the first hours. This indicates an induction period. At 270°C, the induction period is shorter. Asphalts blown at 250°C contain less resins and more polycyclic aromatics as compared to those blown at 270°C. The reason for this is that the conversion of resins into asphaltenes takes place more rapidly, than the change of aromatic compounds into resins.

The oxidation of a steam reduced crude oil and a cracked petroleum residue by air and oxygen blowing at 200°C has been investigated by Graham, Cudmore, and Heyding[15]. The change in the quantities of oils, resins, and asphaltenes with oxidation has been determined, as well as the oxygen content of these fractions. The oxygen absorbed by the residue was found to be small compared with the oxygen consumed in the formation of water and carbon dioxide. The heat evolved during blowing asphalt was measured by Smith and Schweyer[16], and its calculation was also taken up by them.

One cannot draw the conclusion that the heat of reaction per mole of reacting species varies directly with these data, because the change in Ring and Ball softening point is not necessarily linearly related to the change in chemical properties. Consequently the authors are presently considering an experimental method to measure the heat of reaction per mole of reacting species.

2.5.1.3 *Reaction Kinetics*

Considerable difficulties arise in the kinetic treatment of the asphalt blowing process especially in its mathematical approach as a consequence of the following circumstances:

a) Material transfer surface changes during reaction time, i.e. existence of bubbles, by bubble coalescence (outrunning phenomena, bubble increase caused by static pressure decrease, or gas temperature increase),
b) O_2 absorption from a closed system (air bubble), producing a concentration gradient decreasing with time at the boundary surface for the reacting O_2,
c) Diminution of partial pressure of O_2 owing to the decrease of static pressure during the ascension of the bubbles,
d) Diffusion of a chemically inert N_2 front through the boundary layer into the feed due to O_2 absorption,
e) Change of the diffusion constant during blowing as a function of the viscosity (temperature) of the reagents.

The notion of the reaction rate which is very important in kinetic treatments has not yet been cleared in connection with asphalt blowing.[17]

The blowing product is generally investigated by the increase of Ring and Ball softening point (decrease of the penetration at 25°C, increase of the asphaltene content and viscosity. This method is rather convenient, since it must be carried out anyway for the sake of the quality control of finished products. The actual blowing process is however not followed up exactly by it, and is rendered with great delay (1 to 5 hrs), unless a process viscosimeter is utilized. This is due to the consecutive character of the reaction, the formation of intermediary products which cannot be detected by other methods. Stripping effect and subsequent reaction are measured upon conclusion of the blowing process.

Lockwood[18] stated in the kinetic analysis of blowing that the process takes place as a reaction of first order within certain limits. A special type and reaction order must be selected during kinetic studies, since any number of reactions are involved. The softening point is considered the variable in the design and operation of a plant. The simplest reaction type which may hold good for the batch blowing process is an irreversible primary fractional reaction. Laboratory tests of Holmgren proved that the equation obtained is not valid for small air rates and high temperatures. The Arrhenius graphs plotted on the basis of data published by Rescorla and co-workers show that the reaction rate constants are functions of the mixing velocity and the air rate. It is indicated by this that the blowing reaction is partly governed by diffusion. The overall reaction rate constant may depend on three factors in a heterogeneous reaction like blowing, the liquid diffusion, the chemical reaction, and the gaseous diffusion. The liquid diffusion is of no importance under the prevailing rate of mixing.

The possibility of characterizing the reaction rates by the heat of reaction/time unit during blowing is indicated by Senolt[(19)] on the basis of a plant test series. A drawback to this method is that stripping effect and after hardening cannot be measured, but it indicates that the reaction rate remains constant over the temperature range from 220 to 320°C, and is only a function of the available material transfer surface, as are all the combustion reactions. The work of a plant vessel was perfectly linear after a short initial period; the temperature increase per unit time of Austrian asphalt from Matzen was the same at 220°C over the whole range up to 320°C, as long as crude oil parts suitable for blowing were not lacking, or no quick coalescence occurred due to high viscosity (high Ring and Ball results).

The occurrence of after-hardening in large plant vessels cooling slowly proves that the actual oxidation reaction to form H_2O and CO_2 does not take place immediately at the boundary layer, but later in the liquid phase. After all, this fact renders the blowing process feasible, since otherwise the high temperatures caused by oxygen quantities remaining at the interface asphalt/air would immediately result in spontaneous combustion.

Apart from the produced asphaltene types, ester or C-C asphaltenes, stripping is also very important for the reaction products, since it is not the same whether a given Ring and Ball penetration is achieved by asphaltene content inrease, or decrease of oily constituents in the feed stock.

Since the bulk of the above described phenomena occur only in commercial scale vessels, greatest care must be taken should laboratory investigations be applied in practice. Strictly speaking, laboratory tests are suitable only for the comparison of various feed stocks.

The investigations on the blowing process carried out in the last decade resulted in a great many valuable findings. It should be advisable to obtain comparable result to analyse various feed stocks blown under different conditions by the same method. It is desirable that in the course of further investigations this variety still causing disturbances will be eliminated and a thorough knowledge of the influence of the oxygen in the air on the asphalt structure will be determined.

2.5.2 Laboratory Tests

The knowledge of the blowing process, and that of the factors bearing on the quality of the finished products must be assumed when resorting to blowing asphalts. The results of the research carried out systematically in this field were made known mostly in the last decade. Studies were effected in the laboratory and pilot plant, with the purpose of improving the blowing apparatus technology developed as a result of the application of recent data, and also of obtaining, suitable data for the solution of design problems.

Comparative studies were made in a unit with or without mixing by Rescorla, Forney, Blakey, and Frino[(20)]. The ratio of liquid level to diameter was 1:1 in the blowing apparatus. On the basis of previous investigations, this ratio seemed to be best when utilising a mixer. Experimental results indicated that blowing time is shortened by the application of a mixer. Further improvement of efficiency could be achieved by a second mixer supplied in the same vessel or in a connected vessel, as was shown by determining the oxygen content of exhaust gases. The height to diameter ratio was changed within wide ranges in blowing condition studies utilising the test apparatus of Chelton, Traxler, and Romberg[(21)]. Increasing the height of the liquid column ensures quicker increase in the softening point without changing the softening point-penetration ratio of blown asphalt. Test results call for the application of vertical blowing apparatus. By increasing air velocity from the conventional values up to its tenfold, blowing time is shortened considerably, however the penetration of an asphalt of similar softening point but obtained at higher air velocities is lower than that of asphalt made at low air velocities. Blowing velocity increases also with pressure increase.

Blowing temperature and air consumption were investigated in a test series by Gun and Gurevitch[(33)]. Studies made on Baku residue indicated that blowing time and air consumption was at a minimum at 250°C, and reached the highest value at 210°C. Air consumption increases considerably when the temperature rises from 250°C to 270°C. This statement is in agreement with the conclusions drawn by Goppel and Knoterus. The distillate was completely recycled as a reflux in the test apparatus of Murayama, Fukushima, Fukuda, and Shimada[23]. In this case, the penetration, the softening point, and the ductility of the blown product are determined by the feed stock and are not influenced by blowing. Direct water pumped is applied in their test apparatus to diminish the reaction temperature. The softening point-penetration ratio of the blown product was not altered by this cooling method. Gundermann and Müller[24,25] used a testing equipment by means of which the liquid height of the material to be blown and the air distribution could be changed. An internal mixing device was used in part of the tests. Experiments in an apparatus with simple air distribution resulted in raising the blowing effects by increasing the filling height and thus lengthening the way of the gas bubbles. However, a certain height, which has to be determined in each case, must not be exceeded, since then no increase takes place. Again, increasing

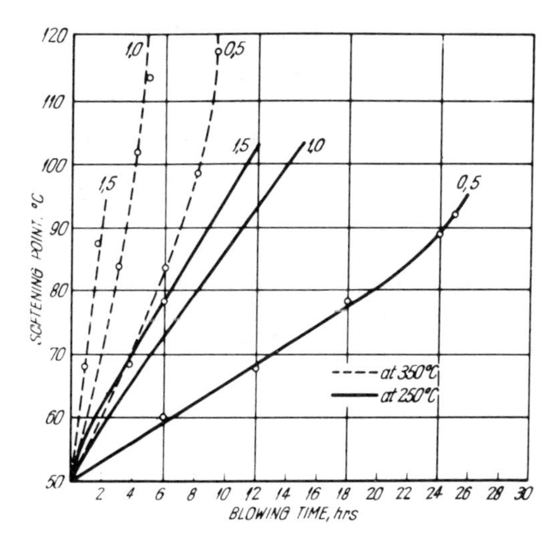

Fig. 24. Raising of softening point of Nagylengyel asphalt (s.p. 50°C) as a function of blowing time, temperature and air velocity. The figures at the curves indicate air velocities in cm^3/min g

the liquid height produces no effect in a blowing equipment with good air distribution and predominantly turbulent flow, since the best use is made of air even at relatively slight liquid heights in a similar apparatus. Increasing air quantities under equal conditions results in a greater blowing efficiency. As regards the effect of larger air quantities on the reaction time, direct proportionality is experienced in all cases. The more or less pronounced effect of larger air quantities seems to depend also on the apparatus and the nature of the asphaltic feed.

The effect of air velocity, blowing temperature, and feed stock quality was investigated by Zakar and Mózes[26] for their influence on blowing. Feed stocks of different consistency and various origin, and those of similar origin but different consistency were investigated. The blowing velocity also rises, when air velocity is increased. Correlations of data obtained in the investigations at 250 and 350°C are shown in Figure 24.

The softening point-penetration correlation of the asphalts studied which were made at a space velocity of 0.5, 1.0, and 1.6 cm^3/min g, as well as other asphalt properties seem to be practically independent of the applied air velocities. The blowing velocity increases when the temperature rises. The blowing relationships of various feed stocks are given in Table 23.

Increasing blowing temperature diminishes the penetration of blown asphalt and increases the breaking point as well as the asphaltene content of the asphalt made. Changes in asphaltene content as a function of the blowing time characterized by the softening

TABLE 23.
Reaction Velocity at Various Temperatures

Feed stock	*Average softening point increase °C/hr at °C* 250	300	350
Residue (Nagylengyel) 20°E at 100°C	5.0	11.5	25.0
Asphalt (Nagylengyel) Soft. point 50°C	3.5	7.5	13.0
Asphalt (Nagylengyel) Soft. point 89°C	1.5	—	8.5
Asphalt (Lispe) Soft. point 52°C	—	—	14.0
Asphalt (Matzen) Soft. point 52°C	1.6	4.0	10.0

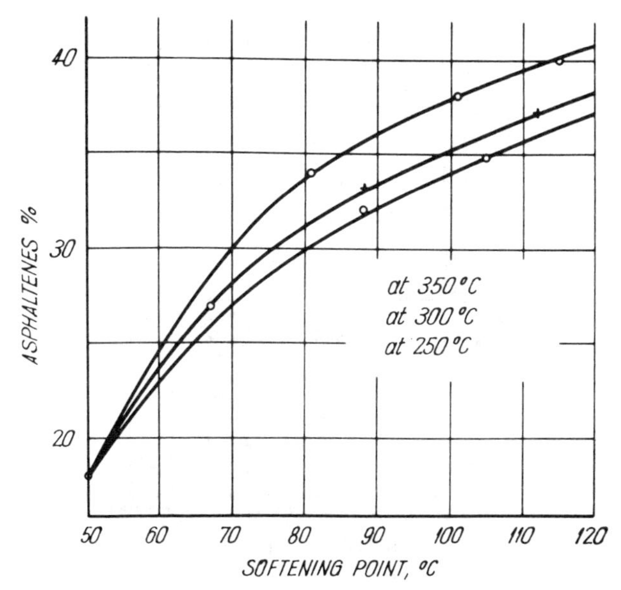

Fig. 25. Changes in asphaltens contents at various temperatures. Feed stock is Nagylengyel asphalt (s.p. 50°C)

point are shown for various temperatures in Figure 25. The blowing velocities of feeds of different origin and the same consistency are different.

Depending on the origin and consistency of the feed stock, the quality of blown asphalts changes, the softening point-penetration ratio.

When blowing asphalts with the same softening points and penetration, but of various origin, asphalts with different penetrations are made when the softening point is identical. Figure 26 illustrates the softening point-penetration correlation of blown asphalts made from feed stocks of the same origin, but of different consistency.

When blowing asphalts on the basis of application of test results it must be considered that the blowing expenditure of the process is considerably increased if the air velocity is raised due to the required larger compressor capacity and electrical power. The increase of blowing velocity at a higher temperature influences the quality to a great extent, and is therefore limited by the quality of asphalts to be produced. The selection of the most suitable feed stock depends on the existing possibilities and the desired quality of the asphalt. The blowing velocity is also affected by them.

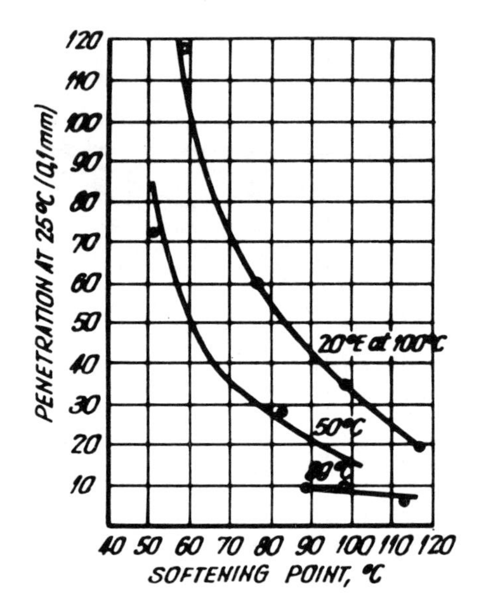

Fig. 26. Correlation between the softening point and penetration of blown asphalt made from feed stocks of various consistency. The figures at the curves represent softening points of feed stock in °C

A new film process of asphalt oxidation is communicated by Provinteev[27], resulting in the method of emulsion oxydation of asphalt.

An equipment was developed by Klimke, Mothes, and Kohlstrung[28,29] consisting of an electrically heated double cylinder. It has also two zones, and the distillation residue preheated to 60°C enters the outer zone from the bottom. The product is heated to the determined reaction temperature during ascension, and mixed thoroughly with injected hot air. The product moves in this zone in the shape of a film having a thickness of a few centimetres and descends from the top of the vessel again in the inner zone as a very thin film, similar to that used in thin film distillation. The reaction is finished in the inner zone, and good separation of the blown oils is achieved, which are driven off as an overhead with the excess air and condensed there. A pilot plant was erected too.

Units are described by Mothes, Prinzler, and Klimke[30], in connection with the pilot plants for Romashkino asphalts. Higher efficiencies are obtained with these units than with the above mentioned ones. Combined compact thin film blowing, blowing in packed towers, and cascade blowing are taken up. Experiments indicated that the cascade principle is best for large scale plants, corresponding to the present technical level.

A new type vessel with a circulation system is related by Mózes, Kádár, and Kristóf[31], which was used in pilot plant experiments based on laboratory test results. The principal parts of the equipment are two concentric tubes, the foam breaker, the upper part to take off the products and vapors, as well as the blower system mounted at the bottom.

Several other solutions are described in the literature[32,33]. A good survey was compiled in this field[34]. However, no particulars have been given on the application techniques of these methods up to now. Finally, it can be stated that operating conditions and the construction of blowing equipment will be greatly influenced by the blowing process and the quality of the blown asphalt depending on feed origin and quality. Optimum conditions of blowing are to be determined considering the desired purposes and the feed stocks available. No doubt a systematic test series is required for this. The development of blowing equipment with higher efficiency must be further studied in future.

2.5.3 The Blowing Process in Industry

2.5.3.1 Techniques for the Manufacture of Blown Asphalts

Various equipment is used commonly for blowing asphalts. In the individual units, there are horizontal or vertical apparatuses, used,—operated batchwise or continually. Great differences are found also in the method of air supply and in the treatment of exit gas. The earlier blowing units were operated batchwise with horizontally arranged equipment (Fig. 27). The asphalt to be blown is heated by outside heater to the initial blowing temperature. Air distributed by perforated coils was pressed through the material to be blown in the lower areas of the still. Gases evolved in blowing leave through lines of large diameters. A heating coil and a cooling coil, are fitted

Fig. 27. Horizontal blower

in the lower area of the vessel in some units[35]. When blowing begins, heat supply is shut off and later cooling may be resorted to. Blowing equipment was also used in which several vessels were arranged in series. Flowing through these, asphalt is blown continuously. However, blowing time had to be lengthened in equipments with horizontal stills due to the poor air distribution, and the products made did not meet all the requirements. Blowing units operated by this system are now used only in certain cases.

Due to increases in asphalt production, the manufacturers were induced to develop a better process to avoid the long blowing times experienced in units with horizontal stills. Any number of recommendations for the improvement of the blowing process has been made, as can be seen in the patent literature, but only a few could be actually applied in industry. In the first place, differences arise as to the more intimate contact between asphalt and air as well as concerning the continuous operation. Recommendations for the recovery and the disposal, respectively, of spent gases and distillate also differ. In the course of industrial development, horizontal stills from 4 to 10 cubic meters used in the beginning were replaced by vertical vessels which were considerably larger and thus improved results could be achieved.

A blowing still supplied with internal mechanical agitation, an electrically operated Turbo-Mixer to ensure better air distribution is described by Holland[36]. The air enters the still from the side. Air distribution is intensified by baffles opposite the air inlet in the middle and by mixing plates fixed on the impeller shaft. A new system was developed by arranging such stills in series. The feed stock is pumped to the first still and from the top of this enters the bottom of the second still, and it is fed into the third still then in a similar way. To ensure the flow through the individual stills, several connecting coils are supplied for the suitable level setting of the asphalt in the individual stills. The air required for oxydation is pressed separately into the bottom of each still during blowing. The gases formed in blowing escape through a common line direct into the scrubber. The flow sheet of this process is shown in Figure 28.

A continuously operated blowing apparatus consisting of eight vessels is described by Mihailov[37]. The vertical stills are mounted on sockets and there is a level grading between the stills. The first still is the feed tank from which the material is pumped into the next still, which is the still proper. The material flows from this into the following stills as a result of gravity. The air necessary for blowing is fed

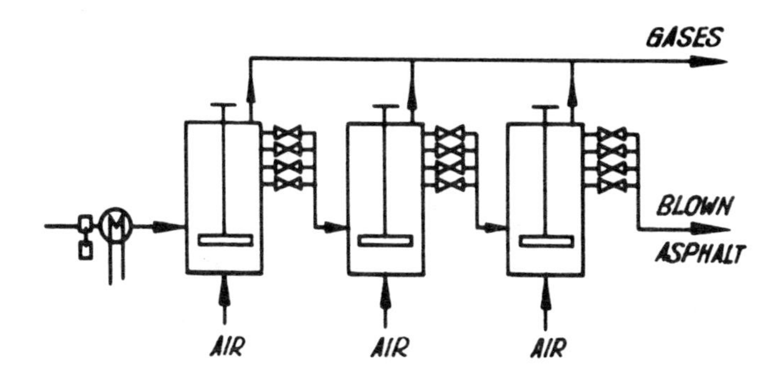

Fig. 28. Blowing stills supplied with internal mixers

into the stills separately, and strictly controlled. The finished asphalt is removed by pumping from the two last vessels.

A one-stage vertical blowing equipment is described by Jackson[38], by means of which continuous blowing may be carried out. The capacity of such vertical towers is from 20 to 80 cubic meters, as stated by Hoiberg[39]. According to Kirk, Othmer[40], a system is developed consisting of such vertical columns where the columns are connected to the distillation unit enabling the feed stock to enter one of the columns, while in the other column blowing can take place. Beside batch processes, continuous blowing may also be effected in these units, when asphalt is introduced overhead into the column and the blown material discharged at the bottom. Column volume as well as those of the vessels required must be suitably adjusted to control operation and blowing times correspondingly between the separate periods of the process. A continuous blowing unit is given by Uhl[41]. It is blown through a carefully designed jet system developed by the Turbo Mixer Corp. The feed stock is pressed direct from the vacuum tower by means of a feeder into the system consisting of two stills arranged in series, after having been cooled in an intermediate cooler to the desired temperature. The height of the asphalt in the still exceeds the still diameter only slightly. The asphalt fed to the bottom of the first still flows through the line after air being blown into the upper area of the still and then to the next still. Air from the feed pipes impinges on the target and enters the impeller between the flowing streams of asphalt. As the asphalt and air mixture is discharged into the stationary ring blades, the air is broken into fine bubbles and violently scrubbed

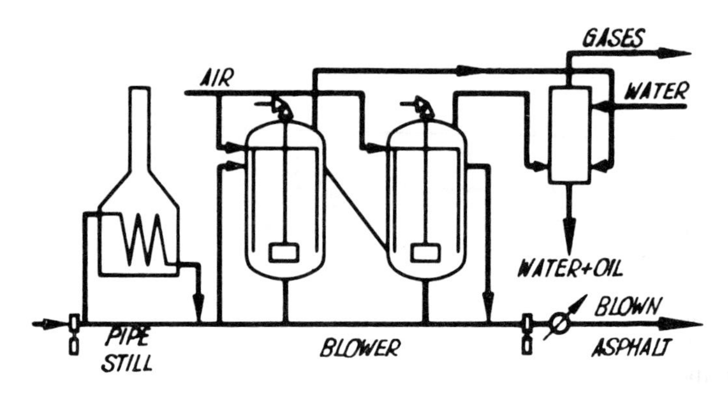

Fig. 29. Continous blowing unit

into the asphalt. In the opinion of plant management, a favorable effect results. A similar continuous blowing unit is discribed by Aixinger[42]. It consists of two stills of 50 cubic meters each. Air is introduced by four tubes with 100 mm diameter each in the individual vessels. Air distribution is effected by an internal mechanical agitation device. A separate vertical tube still is included in the system[43] to heat the asphalt enabling independent operation of the blowing equipment of the distillation unit. The flow sheet is shown in Figure 29. The asphalt unit proposed by Fan-Yun-Nan[44] consists of three vertical towers. The blowing process proper is carried out in the first, largest tower. This tower is equipped with an internal water cooler to control the tower temperature adequately. Two smaller towers serve to stabilize the blown asphalts. To ensure material flow from the individual towers, the inlet point to the next tower is always placed somewhat lower than that of the previous tower; air required for blowing is pressed into the bottom of each tower.

As for these processes, it must be stated that methods operating with internal air mix distribution, although having theoretical advantages, have not given comparable results up to now. That is why this process is not yet commonly used[45]. The capacities of continuous systems equipped with vertical blowing units are growing in industry. Nowadays capacities in some towers operated are 160 m^3 and above. Figure 30 illustrates the flow sheet of continuous blowing equipment used in the asphalt industry now. The asphalt to be treated is pumped from a heated tank through a heat exchanger into the pipe still. The material is

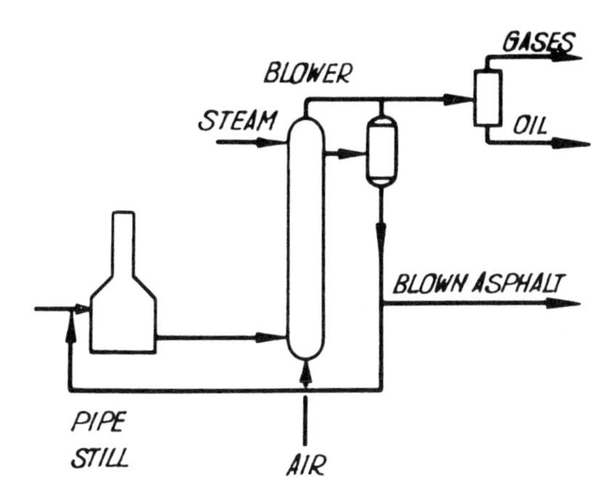

Fig. 30. Continous blowing tower system

heated by this to the temperature required for blowing. Then the material is charged into the vertical tower by a pump.

Two thirds of the tower are filled with the feed stock. Air is blown to the material by an air distributor set at the bottom of the tower. The blown asphalt leaves through the level control tank connected to the top of the tower. The constant rate of circulation passing the heater and the column is operated by the level controller. The discharged product may be recirculated into the system partly or completely through a line connected with the blowing apparatus, if the necessity arises.

Gases and vapors formed during the blowing process leave overhead. Oily distillates are condensed from these by cooling and lead into a tank, while the gases are vented[46]

Beside the above discussed continuous blowing process, blowing can also be made batchwise in units equipped similarly. In some cases batch methods are preferred to those operating continuously.

Continuous processes are most advantageous when large quantities of blown asphalts with definite qualities have to be manufactured from identical feed stocks. The continuous blowing process is very suitable also in cases when the blowing apparatus can be fed from the distillation tower with asphalts of stable qualities. Batch methods are more advantageous in units where different quality blown asphalts are to be made in relatively smaller quantities and the feed stock for blowing must be altered. A drawback to the batch method as compared with continuous blowing is that the treating time is considerably longer.

Several towers are often erected in the plants due to these different viewpoints. These are in general of the same size. However, towers of different capacities are also operated in some plants at the same time[47] with several smaller batch operated towers, whereas the larger towers work by continuous methods. This arrangement permits manufacture of all kinds of asphalt qualities simultaneously.

2.5.3.2 Operation of Blowing Units

The blowing temperature, a function of the feed stock to be treated, of quality and other requirements, as well as of the equipment available, varies between 200 and 320°C. As can be seen from the above details, the material is charged direct from the bottom of the distillation tower into the blowing apparatus, or after passing an intermediate cooler in some of the blowing units. However, this can be made only under favorable conditions, since the bottoms from the distillation tower must have definite qualities to be suitable for the production of blown asphalts.

This process is used first of all where the distillate residue is blown only to a small extent, as in the manufacture of certain paving asphalts. Should asphalts of other qualities be made, operations cannot be conducted by this method. The blowing unit must be equipped with a pipe still to produce asphalts of various qualities and to utilize plant capacities adequately. Various feed stocks are required for the manufacture of different quality blown asphalts. It is advisable to pump the stocks from the feeder tank, heat them to the necessary temperature in the pipe still, and then charge them into the blowing tower. The vertical pipe still proved satisfactory in modern plants for the heating of asphalts. Putting the pipe still in operation, the run is begun with a low viscosity oil distillate or with fuel oil, and only then will the asphaltic material proper be charged by starting its injection into the still. Removal of the heat evolved in the still is to be provided for, since blowing is an exothermic process. The simplest method to avoid overheating is the reduction of the air feed. Apart from this, various direct and indirect methods are applied for removal of the heat produced in the blowing process in plant practice. Part of the vessel surface is cooled by water spray in some units, whereas a steam blanket is used in other cases. Asphalt temperature is decreased by gas oil circulation in certain plants, and internal water coolers are also used for this purpose. Recently, new methods were developed by which water is introduced in the air lines of the blowing apparatus utilizing a feed pump, thus controlling the quality of blown asphalt. The relatively small water quantities are under strict control.[48] Water is sprayed on the surface of the blown asphalt into the gas zone in some units as soon as blowing is finished, and the asphalt is cooled by this to 230°C, as foaming would result at a lower temperature. The temperature of the charged material must be maintained constant when operating by the continuous blowing method, enabling the setting of the desired temperature in the still. The temperature of the feed stock must be maintained lower that of the material in the still.

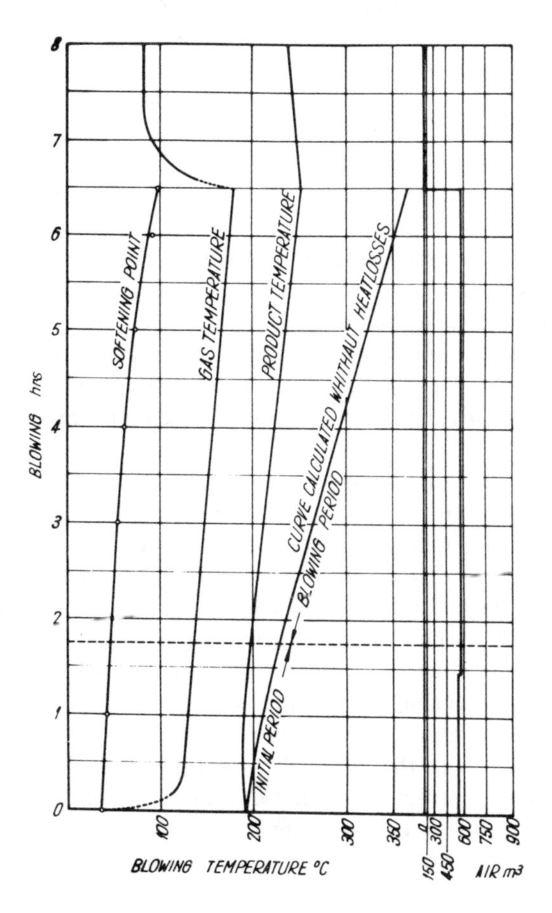

Fig. 31. Temperature vs. time curves of an asphalt blowing process

Curves of temperature vs. time of an asphalt blowing process are shown in Figure 31. Data relate to the blowing of an Austrian asphalt. The Figure also comprises the curve calculated on the basis of data furnished by Senolt without heat losses.

Although asphalt blowing is a simple process in principle, it requires careful control, especially when large quantities are being treated. Flashing can take place due to heat evolved in blowing. A steam blanket is introduced continually in the upper gas space of the tower for precaution. There is another steam line in the lower part of the tower. At times steam is introduced into the asphalt by the steam inlet mounted at the tower bottom, which is to serve the purpose of quality improvement. In any case a certain cooling is arrived at. The blowing unit lines are equipped with additional heating and they must be cleaned by air injection when the unit is shut down.

The air consumed in blowing amounts to 0.5-1.5 m^3/min. t, in general[44]. Air suuply can be provided by a variety of blowers or compressors. The most common is the rotary type. The term blower generally applies to rotary units operating below 0.1 atm. outlet pressure, while the term compressor applies to those capable of higher pressures. Blowers are more economical to operate, but are limited to applications where the operating pressure due to the liquid head in the blowing vessel is within that of the blower capacity. For this the optimum height of asphalt above the air coils is about 16 ft. For greater heights in blowing, use must be made of the high pressure blower or compressor.

Simple low pressure compressors could be used in the old horizontal vessels since liquid level was low. Recent vertical towers require compressors of larger capacities. Since the greatest expenditure is required by the operation of this compressor, its capacity is not raised above a certain limit. To ensure regular air supply, a buffer tank is connected after the compressor and a back-pressure valve is incorporated into the line, otherwise asphalt will enter the air line in the case of breakdown. According to the different constructions as described above, air enters from the tower bottom the vessel in various directions, or is contacted with the material to be blown by a distributing line set in the tower bottom. In other methods, air is introduced from the upper part of the tower bypassing a distributor and passes the coils immersed in the asphalt as far as the bottom of the tower, where it contacts the asphalt in a preheated state. Air distributing systems must be constructed to enable satisfactory cleaning. The air not used during blowing, the gases and vapors formed are vented at the top as overhead by the gas line mounted there. First the nitrogen of air escapes, then steam and the carbon dioxide formed during blowing, as well as the small part of hydrocarbons produced[50]. The vaporized compounds account for the disagreable odor found in blowing fumes. Due to the high oxygen content of the exit gases, coke deposit results in the steam space and in the lines. These in turn are caused by combustion and constrict the lines, involving possible breakdowns. The lines should therefore be constructed in a manner enabling easy disconnecting and cleaning, if necessary. When the total oxygen is not consumed completely, the amount of distillates will increase. Raising the blowing temperature improves the oxygen consumption, but the distillate quantities increase.

A compromise solution is unavoidable here, since explosion may take place due to the high oxygen content of exit gases. Temperature decrease is sometimes resorted to in order to avoid coke formation. However, this results in increased explosion danger because the oxygen content of exit gases increases, and their hydrocarbon content is diminished. These contradictions render the consideration of the explosive limits for the gases absolutely necessary. The explosion limits are illustrated in a triangle graph[51] for the mixture oxygen, nitrogen and oil. The gases escaping from the upper zone of the vessel are lead through various condensers. A knock-out drum or vapor trap in the effluent level after the blowing vessel will help in eliminating these fumes.

Gases are usually contacted direct with water in these condensers to remove toxic and smelly materials, as well as small oil distillate quantities. The water quantity is set in the scrubber so as to condense all the oil vapors present in the gases. The bulk of the smelly material may be removed in this manner. An emulsion is formed between oil and water in the course of this process. It can be broken by precipitation or by other method. Water cooled condensers are used in other processes. Water is separated also here simultaneously with the oil distillates, although to a smaller extent. This has to be isolated afterwards. As a result, an oil distillate is obtained which can be used after mixing with industrial fuel oil, or as a cracking feed. Recovery of the oil is not always economical, In some units the exit gases are vented or introduced into a combustion system arranged for this purpose and ignited. All the smelly materials can be removed by this process, investment and operational costs are however higher than those with

TABLE 24.
Average Distillates (Blown Oils) Obtained from Nagylengyel Asphalt

Specific gravity at 20°C, g/ml	0.948
Viscosity at 20°C, cSt	99.8
°E	13.1
at 50°C, cSt	17.74
°E	2.6
Pour point, °C	± 1
Flash point, Marcusson °C	148
Acid value, mg KOH/g	1.6
Sulfur, %	2.46

water wash, in the case of which spent gases are vented by a stack of suitable height[52].

The condensed product obtained during blowing, the distillate, is an oily liquid containing water, which is a pastelike substance at room temperature in the case of highly paraffinic feed stocks. Data of distillates obtained by blowing Nagylengyel asphalt in a given plant are shown in Table 24, whereas Table 25 contains data resulting from the distillation tests of these distillates[53]. The gaseous products formed during the blowing process and not condensed by water cooling, together with process losses, as well as the part consisting of condensed oils are shown in Table 26[54], based on plant data. It is shown by these data how the losses and the distillate quantities increase as the blowing process progresses, with softening point increase. The quan-

TABLE 25.
Distillation Test Data of Blown Oils

Product	*Pressure Torr*	*Temperature °C*	*Yield %*	*Specific gravity at 20°C g/ml*	*Sulfur %*
Destillate 1	760	until 180	0.8	—	1.70
Destillate 2	760	until 270	8.1	0.8803	1.84
Destillate 3	760	until 310	12.4	0.8985	2.64
Destillate 4	100	until 280	17.6	0.9173	2.75
Destillate 5	100	until 295	9.9	0.9275	2.87
Destillate 6	40	until 308	28.2	0.9418	2.89
Destillate 7	3	until 315	10.2	0.9816	2.96
Residue	—	—	9.9*	—	—
Losses	—	—	3.1		

* Softening point, 84°C

TABLE 26.
Shaping of Blown Oils, Gases and Losses in Blowing Various Residues

Softening point of blown asphalt °C	*A* Blown oils %	Gases and losses %	Total %	*B* Blown oils %	Gases and losses %	Total %
50 — 60	1.0	1.0	2	6	1.5	7.5
60 — 70	2.5	1.5	4	9	2.0	11.0
80 — 90	5.0	3.0	8	14	4.5	18.5
120 — 140	10.0	6.0	16	18	11.0	29.0

Remark: The percentage values relate to the quantity of feed stocks used for blowing.

tity of the produced distillate is a function of the consistency of feed used in the blowing process. When blowing feedstocks of the same origin, but of different consistency, the distillate quantity will decrease with increasing softening point. The feed marked by "A" in Table 26 is a Nagylengyel asphalt having a softening point of 55°C, that designed with "B" is an atmospheric residue from Nagylengyel. The correlations mentioned are illustrated well by the comparison of these data.

The distillate quantity increases with increased blowing temperature. The characteristic data, the yields, and distillate quantities of an asphalt obtained by blowing Nagylengyel residue in the laboratory at various temperatures are shown in Table 27. The penetration decrease of asphalts of identical softening points together with distillate quantity increase are shown in the Table[55]. It ought to be summarized that the changes in the values obtained in plant or laboratory blowing

TABLE 27.
The Influence of Temperature on the Quantity of the Blown Product (Feed Stock; Nagylengyel Atmospheric Residue, 20°E at 100°C)

Blowing temperature, °C	*250*	*300*	*350*
Blown asphalt;			
Softening point, °C	98	100	97
Penetration at 25°C, 0.1 mm	35	—	17
Yield, %	83.3	80.0	68.3
Blown oils, gases and losses, %	16.7	20.0	31.7

tests on Nagylengyel asphalt serve only to illustrate the trend in these changes, which are due to different conditions. Blowing identical asphaltic material in different laboratory or plant vessels give various results. Different changes occur in blowing asphalts of different origin depending on their chemical constitution.[56]

To achieve the strict control of the blowing process, the equipment must be supplied with measurement and control devices in an adequate manner. The control and recording of tower temperature is of outstanding importance apart from the measurement of air quantities. Level control and immediate indication of any breakdowns taking place is also vital. First, suitable temperature measurement points must be inserted to solve these problems. Foaming of blown asphalt occurring from time to time due to various circumstances is a well known phenomenon for the staff servicing blowing units. In such cases the gas space over the blown asphalt is filled with the foaming asphalt and an unpleasant foam gush can happen at the safety valve. In some cases, especially at the end of the blowing process, coke deposits are ignited in the gas space of the still due to the intensive oxidation, and serious fire danger may result from the overheating in the upper space of the still.

According to Kostrin[57], a thermocouple should be incorporated also in the gas space of the still as a precautionary measure and to eliminate these undesirable phenomena. This enables better control of the processes taking place in the still. The highest temperature measurement point is from 1.2 to 1.5 m above the asphalt level. The temperature of the material to be blown, generally 240-260°C, is shown by the thermocouple fixed below the asphalt level. In that case, the thermocouple placed in the vapor space will show 210-225°C. The temperature difference between the liquid and the gas space has to be stipulated in the design to amount to at least 35°C. If foaming occurred in the still, the thermocouple placed in the vapor space would be reached by the material and the thermocouple temperature would immediately assume the value of the liquid temperature. Excessive oxidation of the vapors or the probability of coke ignition is immediately shown by the temperature controller in the vapor space in a similar way. Should a sudden temperature rise occur at this point, foaming can be eliminated by cutting off the air feed or regulating it, and also measures can be taken to introduce steam into the gas space. Apart from the above mentioned temperature measurement, it has also proved to be advisable in the lower part of the still. Considering these problems, Gun and

Bakutkin[58] communicated experimental results with radioactive isotopes applied to the level measurement of the blown asphalt. The level measurement with radioactive isotopes is as suitable for this purpose as the temperature measurement using a thermocouple. Since the cost is approximately the same, no substantial difference can be established.

The necessary asphalt/air ratio has to be determined first, and then the blowing temperature set in order to control and adjust the blowing process properly. Changes in the properties of blown asphalt in the course of blowing are controlled on the basis of the softening points and penetrations of samples taken from time to time.

The layout of asphalt blowing units is not uniform. The blowing unit is as close as possible to the distillation tower when the feedstock is charged direct from the distillation tower into the blowers. In other cases the blowing unit is erected far from the distillation plant, near the racks and the transport, as well as to loading facilities. It is usual in both cases to heat the asphalt stored in the intermediate tanks in a pipe still. The blowing equipment consists of several towers corresponding to the plant capacity. The asphalt is allowed to cool in the still in batch blowing operations. It is then discharged directly into the railway tank or into the packaging facilities used.

2.5.3.3 Statement of the Capacity

The capacity of a blowing unit operated under given conditions can only be stated correctly in relation with the feedstock to be treated. The blowing time is shown in hours in some publications. However, this is insufficient since apart from plant capacity it depends on the properties of the feed stock to be processed and the quality of the desired asphalt, and is characterized by all of them together, if blowing

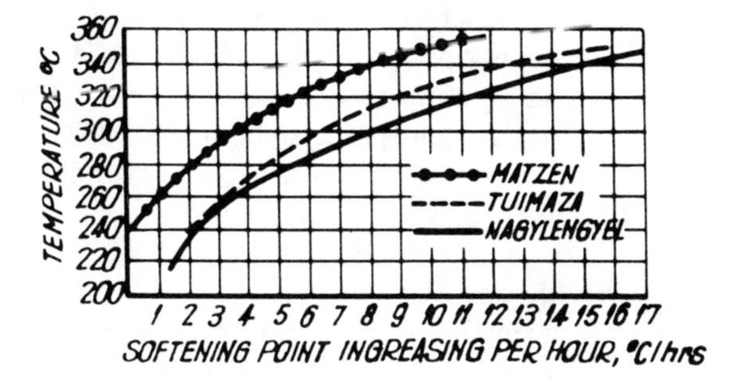

Fig. 32. Softening point increasing per hour of asphalts of various origin having similar softeining points (50°C)

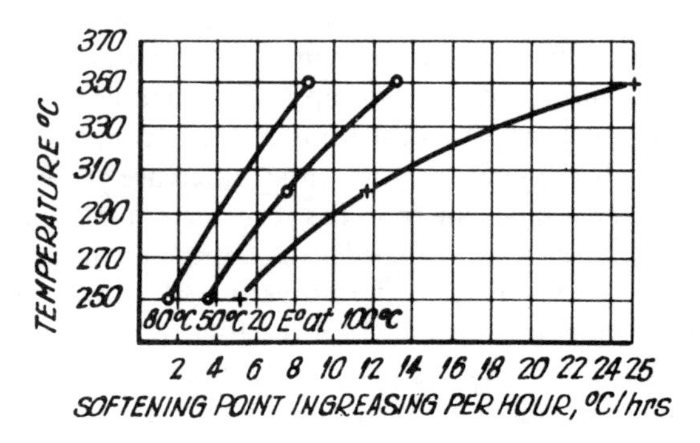

Fig. 33. Softening point increase per hour of Nagylengyel residues with different consistance as a function of the blowing temperature. The figures at the curves represent the softening points of the feed stocks in °C

operation conditions and blowing effect are not considered. The softening point increase per hour of asphalts of various origin having similar softening points under the same blowing conditions is shown in Figure 32, based on the comparative laboratory tests carried out by Mózes and Zakar[59]. It can be seen that the reaction rate is lowest with the feed stock from Matzen, whereas Nagylengyel asphalt exhibits the highest reaction rate. Intermediary values were found in the case of Tuimaza asphalt. Average softening point increases per hour are plotted in Figure 33 of Nagylengyel residues having various consistency as a function of the blowing time. The softer the feed stock, the higher the reaction rate and the greater the blowing velocity change as a function of the temperature. These circumstances must be checked by previous laboratory tests, or better by pilot plant investigations. The desired quality is reached in some plants under the given circumstances by a softening point increase of only 10-15°C. However, in most cases a softening point increase of at least 40°50°C must be brought about by the blowing process. Both the knowledge of this condition, and that of the feed stock are necessary for correct evaluation of plant capacity.

2.5.4 Catalytic Blowing Process

Continuous experiments were made and a great number of catalysts were recommended to shorten the blowing reaction[60] ever since the introduction of the asphalt blowing process. Apart from the action on the blowing rate, an influence on asphalt properties by the catalyst was also found. The experiences of the last decades will be summarized. It was stressed by Pauer and Harumi[61] that mostly undesirable byproducts result when using the catalysts recommended. The removal of the byproducts is cumbersome and expensive. In some cases the catalysts cannot be separated from the finished product at all. When chlorine and its compounds are used, HCl always evolves as a corrosive byproduct. Besides, more or less unstable chlorinated compounds are formed and HCl is split off later, impairing product quality. These drawbacks also occur if chlorine is not applied in a pure state, but is diluted with some inert material, but they can be avoided when small quantities of chlorine are added to the common blowing air. The blowing process then takes place at a temperature lower than usual; the operation time for asphalt blowing is shortened substantially, no halogen substitution worth mentioning occurs in the asphalt, corrosion damages are avoided, and slop oil formation decreases.

Shearon and Hoiberg[62] discuss a catalytic blowing unit in which ferric chloride has been used for ten years, and give full detail of a blower operated by them with phosphorus pentoxide as a catalyst. The same apparatus is used with or without catalyst. The most suitable compound proved to be the phosphorus pentoxide. This catalyst is used in quantities of from 0.1 to 0.3% depending on the desired properties of the blown asphalt. The catalyst is added in the form of a white powder into a vessel which contains the asphaltic material previously charged and heated to a temperature at which it can be mixed easily. The catalyst material required for blowing is withdrawn from the mixture and fed into the retort. To avoid settling, splatter plates are used or the feeding must take place at a point where good mixing is ensured by constant turbulence. The catalyst remains in suspension during the total blowing time by suitable control of the blowing air. While air blowing proceeds, discrete particles of phosphorus pentoxide disappear, and it is thought that an intermediate compound is formed. This intermediate compound may be an ester, an addition product, or perhaps some water resistant complex. Whatever it is, it prohibits the recovery of the catalyst by water extraction of a solvent solution of the blown asphalt, as can be done in the case of ferric chloride.

The time of blowing is usually shortened to a small extent by the catalyst, but it is even lengthened at times. The importance of the process is the possiblity of making special asphalts. Asphalts of 85, 132, even of 176 penetrations can be manufactured instead of a 34 penetration asphalt at 25°C with a given softening point by increasing the catalyst quantity. Catalytic asphalts made in this manner exhibit improved quality as regards the other properties, and are especially advantageous at low temperatures. The phosphorus pentoxide constituent cannot be detected in the blown asphalt. Some researchers think that an intermediary product is formed. Corresponding to this, the catalyst cannot be extracted with water as is the case with ferric chloride. The catalytic blowing experiments of Scheianu, Grigorescu, and Vasiliu[63] had the object of shortening blowing time in the first place. Ferric chloride served as a catalyst. As a result of this investigation, the blowing time was shortened considerably with the application of approximately 0.4% ferric choride.

According to Mapstone and Carius[64], the phosphorus pentoxide effect was variable and of no importance. Again, the blowing time could be shortened by half when 0.7% ferric chloride was used. Zakar, Csikós, Mózes, and Kristóf[65] describe the catalytic blowing of Nagylengyel residues of various consistency. It was stated by them that utilization of ferric chloride catalyst results in shortening the time of blowing considerably. At the same time the penetration of the blown asphalt is increased and the Fraass breaking point is somewhat improved too. Also from a qualitative viewpoint, blowing seems to be more advantageous at 250°C since the catalytic effect becomes rather small at 300°C. The reaction promoting effect by the temperature increase is lower with aluminum chloride than with ferric chloride. However, the properties of the blown asphalt are changed in the presence of aluminum chloride to somewhat greater extent. Characteris-

TABLE 28.

The Effect of the Catalyst and the Catalytic Blown Asphalts (Nagylengyel Feed Stock)

Feed stock	*Distillation residue, 43°E at 100°C*		*Asphalt, Soft. point, 42°C*	
Catalyst	*without*	*with 0.5% $FeCl_3 \cdot 6H_2O$*	*without*	*with 0.5% $FaCl_3 \cdot 6H_2O$*
Blowing time, hrs	21	6	17	5
Softening point, °C	80	87′	85	85
Penetration at 25°C, 0.1 mm	50	54	26	36
Ductility at 25°C, cm	5	3	6	5
Fraass breaking point, °C	−17	−28	−11	−13
Asphaltenes (Hexan), %	32	34	33	33

TABLE 29.
Catalytically Blown Romashkino Asphalt (Vacuum Residue, Softening Point 44°C)

	without catalyst	*with catalyst/$FeCl_3 \cdot 6H_2O$*			
		0.3%		*0.5%*	
Blowing temperature, %	300	250	250	250	250
Blowing time, hrs	12	6	7	4	5
Softening point, °C	84	79	94	76	83
Penetration at 25°C, 0.1 mm	17	24	15	37	23
Ductility at 25°C, cm	35	52	32	67	40
Fraass breaking point, °C	−2	−8	−6	−12	−10
Paraffin wax, %	—	—	—	—	2.4

tic data of catalyst effects and those of catalytic blown asphalts are compiled in Table 28. The shortening of blowing time and the penetration increase are apparent from the data.

The quality of the blown asphalt is influenced by the P_2O_5 catalyst in the first place. Asphalts with high penetrations and low Fraass breaking points can be manufactured with it.

Apart from the shortening of blowing time, it is possible to produce from the residue of lubricating oil of relatively low penetration and low paraffin wax content a 85/25 quality asphalt by catalyst. The results of laboratory investigations made by Csikós, Mózes, and Zakar[66] working on Romashkino asphalt are compiled in Table 29.

Although it is necessary to maintain the paraffin wax content limit of 2% in the case of Romashkino feed stock apart from the other requirements, good results can be achieved according to Csikós, Mózes, and Zakar[67] by fluxing the asphalt feed and by subsequent blowing in the presence of the catalyst. Based on comparative tests carried out by Sole[68] who worked on Iraqui feedstocks, the greatest blowing velocity increase was a chieved with $FeCl_3 \cdot 6H_2O$.

It was shown by investigations of Gundermann[69] that aluminum chloride proved to be the best catalyst at 150°C and for 3 hrs. Medium to hard asphalts with a wide range of plasticity were obtained by his method directly from naphthenic-aromatic soft asphalts.

The effect of sulfur on asphalt quality is also remarkable. Products similar to the blown asphalts may be made with sulfur. Far reaching investigations were carried out in this field to determine the relationships. Among others an extract of solvent lubrication oil refining was treated. Products of higher softening points similar to blown asphalts could be obtained in these tests with the addition of 40% sulfur and by heating to 180°C[70].

The processes taking place on addition of elementary sulfur to asphalt may be divided into two categories. Sulfur reacts with asphalt at higher temperatures chemically to form hydrogen sulphide, whereas the sulfur quantity reacting with the asphalt is slight at lower temperatures. The asphalt is dehydrogenated by the sulfur in the first case. The reaction is similar to that between oxygen and asphalt during the blowing process. Although sulfur reacts with asphalt at a lower temperature than oxygen the treatment with sulfur at high temperatures has not been used yet commercially. Laboratory treatment of asphalts of various origin and 200 penetration with 5% sulfur at 220°C resulted in asphalts of lower penetrations and higher softening points[71]. Approximately 30% of the sulfur used in the procedure combined with the asphalt. The comparison of asphalts of higher softening points produced with sulfur to blown asphalts having the same penetration showed total agreement in the case of Mexico asphalts, however with asphalts of other origin, small deviations were experienced. After these earlier tests, other investigations followed later to clear the influence of sulfur on the asphalt and to study the properties of asphalt produced by this method. Since large sulfur quantities are liberated in petroleum refineries when desulfurizing crude oils with high sulfur contents, this research work seemed to be very promising. The price of sulfur is relatively low, and a process was sought in which product quality would not be impaired and sulfur could be used in the refining of some petroleum products obtained locally. In the experiments of van Ufford and Vlugter[72], 20% sulfur and sulfur containing compounds were added to asphalts from the Middle East and Venezuela at 130 and 110°C. In the beginning sulfur behaves like a plastic component when mixed with asphalt. The mixture has a higher penetration index and a lower Fraass breaking point than the original asphalt.

The plastic properties of sulfur are however lost rapidly, since a transition from a plastic material into a crystalline substance takes place.

If polymethylene tetrasulfide is used instead of elementary sulfur, it is evident that the influence of this polymer on asphalt is very similar to that of sulfur. This mixture exhibits also a higher penetration index and a lower Fraass breaking point. However, the plastic character of the polymer in the mixture is not ruined so rapidly. A disadvantage of the simple organic polysulfides consists in their being not very stable at the high temperatures required for asphalt in normal service conditions.

A great many materials are known from the literature which change the properties of asphalts by reacting with them. Only the sulfuric acid and the acid sludge need to be mentioned having a dehydrogenating influence on asphalt and these are utilized partly in industry on a small scale.

2.5.5 Properties of Blown Asphalt

Besides straight run asphalts, blown asphalts are produced on a large scale. They form approximately from 22 to 25% of the total asphalt world production. Part of this is paving asphalt of lower softening point, but the bulk has a higher softening point and is used in civil engineering, for insulations and other industrial purposes, for flooring and roofing. The penetration limit alone is usually given for blown asphalt produced for paving, as for straight run asphalts. Thus it is somewhat difficult to determine whether a given asphalt is produced by distillation only, or by distillation with subsequent blowing. Based on the investigations of Jurina[73], characteristic data of quality B 200 asphalt are represented in Table 30, manufactured directly from Matzen crude oil, and by the blowing process. On comparison, the somewhat lower Fraass breaking point of the blown asphalt becomes apparent, but it is obvious that if production methods and feedstocks are unknown, it will be impossible to establish whether a paving asphalt characterized by these data was made by blowing or not.

The above example has been cited since blown asphalts are considered unsuitable for paving purposes by several investigators[74]. These statements result from the poor ductility values of blown asphalts. Unfavourable road building experience has no doubt added to this negative viewpoint. Doubts are expressed also due to the poorer

TABLE 30.
Asphalt B 200 Made From Matzen Feed Stock

Production method	*Vacuum distillation*	*Blowing*
Penetration at 25°C, 0.1 mm	201	199
Softening point, °C	39	41
Ductility at 25°C, cm	100	100
Fraass breaking point, °C	−15	−42

adhesive properties. Contrary to these opinions, some paving asphalt is manufactured all over the world by blowing. Such paving asphalts are on the market in the U.S.A., but they are only blown slightly. Some paving asphalts are made by blowing also in the Soviet Union. However, if technical conditions are not strictly adhered to in its production, and the finished product quality is not controlled prior to its exploitation, unfavorable experiences and reactions may be encountered. It is, however, not proved that paving asphalts in the manufacture of which the blowing process was also applied must be considered as low quality as a matter of course.

The purpose of blowing asphalt is to obtain suitable asphalts with higher penetration for various uses. These asphalts exhibit beside their higher softening points favourable rheological properties for various kinds of applications, as shown by standard tests also. Blown asphalts with higher softening points are marketed therefore bearing an indication of the manufacturing method. The indication of the softening point and the penetration is commonly used, and they are necessary to characterize the individual groups, since asphalts with similar softening points but different penetrations can also be produced.

The data of Nagylengyel blown asphalts with similar softening points are compared to those of Nagylengyel straight run asphalts having the same softening points in Table 31.

The higher penetration values and lower breaking points of blown asphalts are strikingly brought out when compared to straight run asphalts, as well as the larger plasticity temperature ranges, lending a rubberlike elasticity to blown asphalts in practice. It is shown by the table at the same time that blown asphalts of various properties can be

TABLE 31.
Blown Asphalts (Nagylengyel) With Identical Softening Points

Production method	*Vacuum distillations*	*Blown* *A*	*B*	*C*
Softening point, °C	87	85	85	87
Penetration at 25°C, 0.1 mm	7	15	25	39
Ductility at 25°C, cm	0	4.5	4.5	4.0
Fraass breaking point, °C	+10	−2	−9	−21
Plasticity temperature range, °C	77	87	94	108

manufactured. When high asphalt content crude oils are to be processed, blown asphalts can be made simply by obtaining distillation residues of suitable consistency. An asphalt of good softening point against penetration value and having the desired qualities can be produced from this residue by the blowing process. The relationship between the softening point and the penetration of blown asphalt from Nagylengyel crude is represented in Figure 34. If this relationship is known, suitable feedstock can be selected to obtain the desired quality.

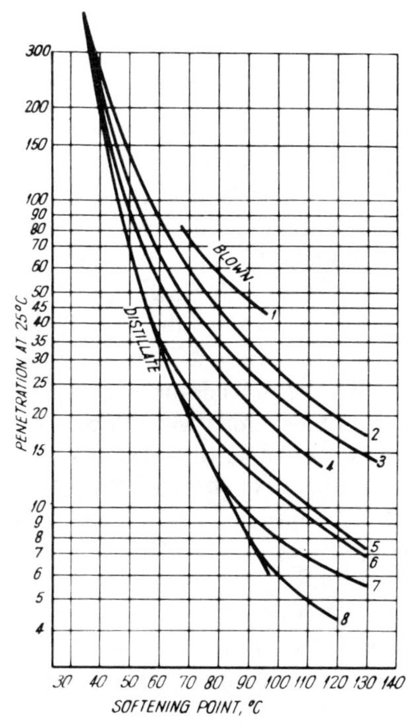

Fig. 34. Softening point vs. penetration of blown asphalt made from Nagylengyel feed

1. from atm. residue
2. from atm. residue
3. from straight run asphalt s.p. 36°C
4. from straight run asphalt s.p. 49°C
5. from straight run asphalt s.p. 60°C
6. from straight run asphalt s.p. 69°C
7. from straight run asphalt s.p. 88°C
8. from straight run asphalt s.p. 92°C

In the case of crude oils used for lubricating oil production, when the asphalt formed does not meet the quality requirements for blowing, a suitably blended feed stock must be made available on the basis of previous evaluation tests. To obtain the desired blown asphalt quality, two residues of different origin and perhaps of different viscosities may be utilized in many cases, or a feedstock composed in some other appropriate manner. Various oil distillates, possibly other residues and solvent extracts should be blended in various proportions with the asphalt at disposal in such cases.

Mixing is done either in the feed stock tank before hand or the components are fed directly into the blowing apparatus in batch blowing; the blowing is first started with a small quantity of air.

The characteristics of asphalts of different origin are represented in Table 32. It is apparent from these data that the properties of blown asphalts vary according to their origin. It should be stressed again, that suitable adjustment of the feed stock for blowing enables the production of the desired qualities in most cases. The problem of origin is also much less important with blown asphalts than with straight run asphalts; this is indicated by the different values of group compositions in blown asphalts. In this connection, Krenkler[75] stated that a straight run asphalt could be judged and evaluated by its origin to a great extent. This should be avoided with blown asphalts. The nature of the applied feed stock is often masked strongly by the chemical reactions accompanying the blowing process. Therefore the question of asphalt origin is of less importance with blown asphalts than that of its actual characteristics.

TABLE 32.
Blown Asphalts Made From Feed Stocks of Various Origin and of Similar Softening Points (From 52 to 55°C)

Origin	*Nagy-lengyel*	*Lispe*	*Matzen*	*Tuimaza*	*Romash-kino*
Softening point, °C	85	82	80	78	77
Penetration at 25°C, 0.1 mm	26	17	11	10	14
Ductility at 25°C, cm	4	2	1	4	4
Fraass breaking point, °C	−11	+3	+5	+5	+1
Asphaltenes (Hexan) %	32	27	17	22	23
Sulfur, %	5.0	0.6	0.5	3.0	3.2
Paraffin wax, %	1.9	4.0	1.2	2.0	2.0

TABLE 33.
Blown Asphalts

Designation	*75/30*	*85/25*	*85/40*	*105/15*	*115/15*	*135/10*
Softening point, °C	80–80	80–90	80–90	100–110	110–120	130–140
Penetration at 25°C 0.1 mm	25–35	20–30	35–45	10–20	10–20	7–12
Ductility at 25°C, cm	4	3	3	2	2	1
Fraass breaking point, °C min.	−20	−10	−20	−8	−10	—
Ash, %, max.	0.5	0.5	0.5	0.5	0.5	0.5
Paraffin wax, %, max.	2.0	2.0	2.0	2.0	2.0	2.0
Flash point, °C, min.	230	240	200	250	250	280
Specific gravity, g/ml at 25°C	1.02–1.05	1.02–1.02	1.02–1.05	1.02–1.05	1,02–1.05	1.02–1.05

The difference in these materials even in the case of identical outside properties becomes apparent by investigating various series of blown asphalts. Even in some of the asphalts with similar properties, great differences in their internal structures could be proved. The qualities marketed in the individual countries for various purposes were partly standardized, but sometimes qualities were specified according to market requirements. The blown asphalt groups and its outstanding quality characteristics[76] as commonly used in the two German States are shown in Table 33. Further brands marketed in the individual countries will be indicated when dealing with various application possibilities.

2.6 Cracked Asphalts

Cracking is a commonly used process in petroleum refining. In the beginning its only purpose was the production of gasoline. During cracking heavy hydrocarbon mixtures, a residue is formed which, however, cannot be used directly as asphalt in most cases. It will often serve only as a feed stock in asphalt manufacture to produce the marketable asphalt qualities by distillation, blowing, and mixing[1].

The temperature commonly used in petroleum distillation does not exceed 400°C in general, since the molecules decompose above this temperature. This phenomenon is used in the cracking process on purpose. The extent of decomposition is a function of the pressure applied and of the time the product is exposed to high temperatures. This temperature amounts to from 450 to 600°C in the usual plant cracking process where gasoline is the desired product. Strict control of plant conditions is of great importance, cracking velocity at 540°C being for example, thirteen times the rate at 480°C.

A heavy asphaltic residue is produced during cracking due to the decomposition of larger hydrocarbon molecules involving dehydrogenation and polymerisation. If this residue is not removed from the system, it is changed into coke. The so-called cracked asphalt during the process is discharged at about 310°C. The asphalt is mildly or strongly cracked in function of the temperature, the pressure, the time of cracking, as well as of the decomposition degree of the molecules. In some cases, the cracking is so strong that the heavy residue can only be used for heating purposes. Sharp differences are exhibited by the residues from the cracking processes in their physical properties and outside appearance, due to the various feed stocks and different cracking conditions. If cracking was conducted with care, the cracked asphalt is as soluble in carbon disulphide as any distillation asphalt. A product being absolutely unsuitable for use can also be obtained owing to the intricate production conditions as mentioned above.

The asphalt produced during the thermal cracking process differs from straight run asphalt especially by higher specific gravity, being hard and brittle, and of higher susceptibility to high temperatures. Its transition from the solid into the liquid state is sharp, therefore it is considered a characteristic pitchlike asphalt. Apart from these properties, its oxidation resistance is lower, which results in rapid damages of the pavements made with it. It is therefore not used in paving or utilized only as a cracked residue blown for the construction of less important roads.

The unsuitability of an asphalt which was cracked to a certain extent during production, was reported by Samokovlija and Grbec[2] based on their comparative investigations. Investigating road sections built with various asphaltic materials, damages soon occurred in the sections constructed with cracked asphalt. Although this asphalt had met the standard requirements, cracking was indicated by the positive Spot-test. In spite of the above statement, there are sometimes reports in the literature on units manufacturing paving asphalts from such feedstocks.

Certain brands of these asphalts are suitable for special purposes. This quantity, however, amounts to only part of the available material.

One of the special applications is briquetting, in which apart from the above mentioned properties of cracked asphalt, its higher Conradson carbon residue proves to be advantageous.[3]

The strongly cracked material is also blended with original material, that is, straight run asphalt. The quality of the material produced in this manner depends on component proportions and the nature of the original asphalt. The miscibility of straight run and cracked asphalts is almost unlimited.

2.7 Comparison of Asphalts Made by Different Methods

The correct selection of the asphalt to be used involves comparing the characteristic properties of asphalts made by various methods.

The characteristic differences become apparent best by comparing the data of asphalts made from the same feed stock, but by various methods. Data on Negylengyel straight run, blown, and propane extract asphalts of similar softening points are shown in Table 34. It is obvious from these data that the penetration of blown asphalt is the highest, and that of propane asphalt the lowest.

Both ductility and Fraass breaking points are highest in blown asphalts. The blown asphalt marked A is made by blowing a softer feed stock, and its Fraass point is therefore lower than that of the material marked B, which was blown from a higher softening point quality feedstock. The asphalt marked C was blown from a 177 penetration (at 25°C) material with 45°C softening point, Ring and Ball, in the presence of P_2O_5 catalyst. The high penetration and low Fraass breaking breaking point of the product are remarkable.

TABLE 34.
Asphalts Produced by Different Methods (Nagylengyel)

Production method	*Vacuum distillation*	*Blowing*			*Propane extraction*
		A	*B*	*C*	
Softening point, °C	98	103	99	98	94
Penetration at 25°C, 0.1 mm	8	22	14	52	2
Ductility at 25°C, cm	0	3	3	30	0
Fraass breaking point °C	+11	−7	+1	−24	+20
Plasticity temperature range, °C	81	110	98	123	74

TABLE 35.
Asphalts Produced by Various Methods (Romashkino)

Production methods	*Vacuum distillation*	*Blowing (atm. residue)*	*Propane extraction*
Softening point, °C	55	56	57
Penetration at 25°C, 0.1 mm	38	68	26
Ductility at 25°C, cm	100	12	100
Fraass breaking point, °C	−6	−24	+1
Paraffin wax, %	2.2	2.5	2.0
Sulfur, %	3.5	3.0	3.6
Plasticity temperature range, °C	61	80	56

The same qualitative differences are exhibited by plasticity/temperature range data also.

Data of Romashkino asphalts produced by various methods and having low softening points are represented in Table 35. It is shown by these data that the comparative statements with regard to Nagylengyel asphalt can be applied to these data also. Paraffin wax and sulfur-contents are shown in this Table as well. With regard to the low ductility of blown asphalt, it must be admitted that the ductility of asphalts blown asphalt, it must be admitted that the ductility of asphalts blown from low viscosity residues with given softening points is very much lower than that of the straight run asphalts, or even that of the propane asphalt. Similar results were obtained in the course of blowing Nagylengyel or Tuimaza residues. Group composition data as for Nagylengyel asphalts represented in the Table are pictured in Figure 35. These render possible the combination of the differences experienced in the standard tests and those of the group composition with asphalts made by different methods. It is shown by group composition data that the propane asphalt has the largest asphaltene content, whereas the smallest asphaltene content was found in distillation asphalt. Blown asphalts possess the highest, propane asphalts the lowest oil contents. Blown asphalts exhibit the lowest resin content. The asphalts made by various production methods possess various properties corresponding to the different structures. Contrary to the brittleness and stiffness of propane extract asphalts, blown asphalts possess certain elastic properties. Straight run asphalts are intermediates between these two groups. The differences between asphalts of different manufacture are experienced naturally in their other characteristics also, which are not discussed here, such as their rheological behavior.

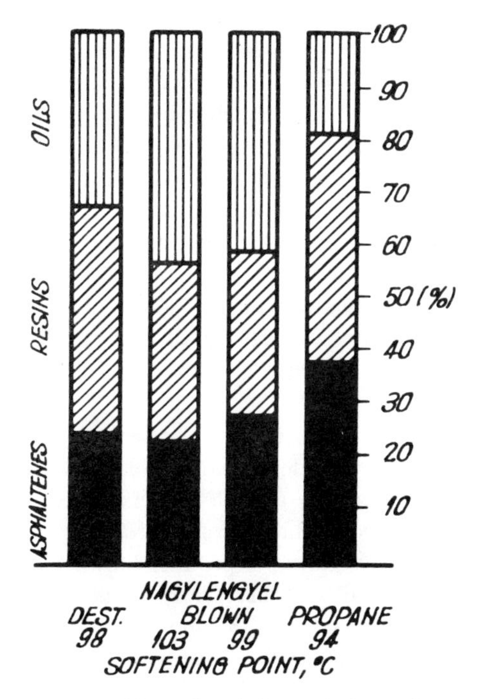

Fig. 35. Group composition of asphalts made by various methods

The range of the differences in asphalt properties and compositions would be even wider, if the comparison covered asphalt groups obtained from cracked feed or catalytically blown material also. Characteristic features of the above have been discussed in the foregoing.

It is apparent from the above statements that a more or less suitable quality asphalt for a given purpose can be obtained from the same feed stock by various production methods. This results in an indication as to which method the best quality asphalt for a given purpose may be made. The characteristic features mentioned of the asphalts made in a given production manner from various feed stocks change as a function of feed stock characteristics within a wide range.

2.8 Other Treatments and Production Methods

The above discussed production methods relate chiefly to the problem of asphalt manufacture from crude oils or process residues. Since the resulting qualities often do not meet the requirements, a secondary treatment also becomes necessary. The blending of asphalts made by various methods with one another or with other materials also belongs to these secondary processes. Production of cut back asphalts are also considered under this heading. Lighter petroleum distillates are mixed during processing with the asphalt. The asphalt emulsion obtained from asphalt and water is made by a special technique and needs to be mentioned here. Numerous production methods for the purpose of making new products by blending the asphalts manufactured according to the above detailed processes with further additives will not be dealt with here. These procedures are used mostly in small units established to meet special requirements.

2.8.1 Blending of Asphalts

The most frequent secondary treatment carried out after the primary production process in the asphalt unit is the blending of asphalt. The process by means of which the feed stock necessary for blowing is rendered suitable by blending to produce blown asphalt of the desired quality will not be dealt with here, only the blending of finished components will be discussed. In the most favorable case, the asphalt manufactured by a given method meets the requirements and can be marketed immediately.

In other cases, however, the produced asphalt must be adjusted to a definite quality by blending. Two or more asphalt qualities of various origin, may be of different consistencies or made by different methods must be blended(1). A commonly used process is to blend the asphalt with a certain amount of another petroleum product to develop a new asphalt quality by suitable selection of the blending component. Apart from the production units constructed specially for the purpose to make desired quality asphalts from various feed stocks, blending may also be blended at the consumers', if the necessity arises.

Penetration vs. softening point curves of blends of straight run 50 and 181 penetration asphalts are shown in Figure 36, based on the investigations of Nyul, Mózes, and Zakar(3) to illustrate the effect of blending straight run asphalts of various penetrations. It is apparent from the Figure that a relatively small change in penetration results from great changes in the blending ratios. No difficulties are encountered in the adjustment to a desired quality. On comparison of the properties of the mixture, it can be stated that an asphalt similar to the straight run asphalt of the same quality can be made by blending

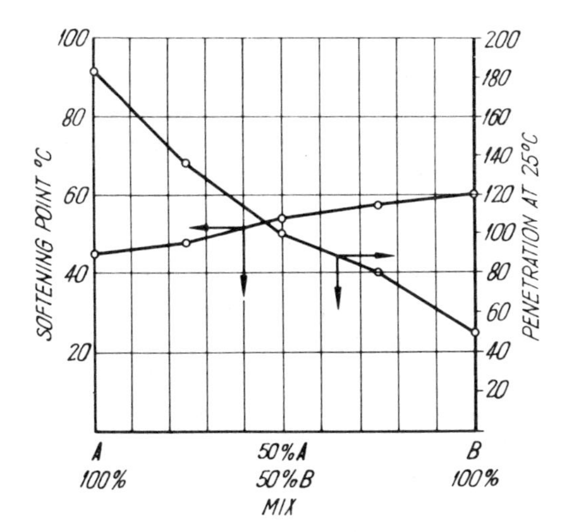

Fig. 36. Penetration vs. softening point curves in blending Nagylengyel straight run asphalts with 50 and 181 penetrations

straight run asphalts. The data of a blend of Nagylengyel asphalt (softening point 75°C) with an atmospheric residue (viscosity 30°E at 100°C) are represented in Table 36 on the basis of studies made by Zakar[3];

According to experiments, the behavior of straight run asphalt blends is similar to that of blends of high vacuum reduced asphalt with blown asphalt. A given kind of asphalt may be obtained by blending a high pour point asphalt with a low pour point quality. In mixing high

TABLE 36.

Asphalt Mix (Blend of Nagylengyel Asphalt s.p. 75°C With an atm. Residue 30°E at 100°C)

	Amount of residue in the mix, %		
	45	*33*	*23*
Softening point, °C	47	54	58
Penetration at 25°C, 0.1 mm	120	75	47
Ductility at 25°C, cm	100	100	100
Fraass breaking point, °C	20	—	—
Loss on heating, %	0	0	0
Softening point of residue, °C	54	60	65
Increase of softening point upon heating, °C	7	6	7
Penetration decrease, % of original	33	40	21
Ductility of residue cm	75	58	32

vacuum reduced-, and straight run asphalts, blown asphalts and straight run asphalts, as well as blown asphalts with high vacuum reduced asphalts, the softening points and penetrations will be lower than the values calculated on the basis of the mixing rule. However, the differences experienced may be neglected from a practical viewpoint[4].

Blending straight run- and blown asphalts of the same origin may be advantageous for other reasons also. According to Senolt[5] this can be used with Matzen asphalt for the purpose of increasing the asphaltene content without changing the penetraction at 25°C and the softening point, Ring and Ball. For example, 100% asphalt blend, mark B 200 with 8,9% asphaltene content can be obtained by blending 85% Matzen straight run asphalt B 200 having an asphaltene content of 2,5% with 15% gel asphalt 85/25 of 25,0% asphaltene content.

The necessity of blending asphalts of various origin with each other often arises. The required softening point/penetration relationship, the adjustment of a given paraffin wax content, the improvement of the breaking point can be achieved well by blending.

The blending tests on Nagylengyel and Romashkino asphalts carried out by Zakar and Mózes[6] are shown in Table 37. Manufacture

TABLE 37.
Blends From Nagylengyel and Romashkino Asphalts

	Nagylengyel asphalt	*Quantity of Romashkino asphalt in the blend %*			*Romashkino asphalt*	*DIN 1965 B 65*
		20	*50*	*70*		
Softening point, °C	50	50	51	51	52	49–54
Penetration at 25°C, 0.1 mm	90	82	69	59	49	50–70
Ductility at 25 °C, cm	100	100	100	100	100	100
Paraffin wax, %	1.8	—	—	—	1.9	2.0
Asphalten (Hexan)	23	21	16	13	7.3	—
Fraass breaking point, °C	−18	−17	−14	−12	−7	−8
Loss on heating, %	0	0	0	0	0	1.0
Softening point of residue, °C	55	54	54	54	56	—
Increase of softening point upon heating, °C	5	4	3	3	4	10
Penetration decrease, % of original	29	32	27	28	29	60
Ductility of residue, cm	100	100	100	100	100	50

TABLE 38.
Blends of Nagylengyel and Matzen Asphalts

	Matzen asphalt	*Quantity of Nagylengyel asphalt in the mixture*			*Nagylengyel asphalt*
		20	*50*	*80*	
Softening point, °C	34	40	55	74	89
Penetration at 25°C, 0.1 mm	318	167	42	18	9
Ductility at 25°C, cm	100*	100	100	11	4
Fraass breaking point, °C	−18	−18	−8	−4	+3
Ashalten, (Heptan) %	0.5	7.2	18.4	27.7	30.0

* at 15°C

of a product corresponding to the quality B 65 by blending serves as an example. Neither Nagylengyel nor Romashkino asphalts meet the standard specifications. The desired quality can be produced by means of blending the two asphalt types.

Nyul, Zakar, and Mózes[7] studied the characteristic data of Matzen asphalt with softening point 34°C, and those of Nagylengyel, softening point 89°C, as well as the characteristics of their blending products. The data are shown in Table 38. Beside standard test data, it also contains the asphaltene contents. Riedl[8] investigated Tuimaza propane asphalt blended with suitable plasticizers. He concluded that adequate paving asphalts with low softening points can be obtained from such blends. Pass and Schindel[9] reported in their paper on Austrian asphalt on asphalts produced by blending straight run- and blown asphalts obtained from Matzen crude oil. The products correspond to the quality specification.

Experimental results are discussed by Sotir and Simionescu[10] obtained in studying paving and industrial asphalts. These were made by suitable blending of propane asphalts and the lubricating extract of furfural refining, and the products blown from these. New techniques in connection with these results are also reported. The correlation between the chemical composition and asphalt properties was studied by Simpson, Griffin, and Miles[11]. They indicate that the most advantageous properties can be developed when optimum asphalt structure is realized. Good results are achieved in plant practice by blending distillate residues of high softening points with suitably selected oil distillates, extracts, or raffinates.

Beneš[12] reports on the production of special asphalts, in the course of which an asphalt with softening point of from 140 to 150°C is obtained which is then mixed with a satisfactory oil to make an asphalt of low breaking point for a special purpose. It is pointed out by Barth[13] that asphalt qualities can be made by blending suitably selected materials which cannot be produced by other methods. The resinous components and asphalt constituents from extraction with propane are good feed stocks to develop synthetic "tailor made" asphalts and ensure extended possiblities also for asphalt manufacturers.

To meet the different requirements and consider the various viewpoints, blending of asphalt must be resorted to as the last process step, which is especially advantageous for the manufacturer in the case of variable feed stock supplies, and ensure the satisfaction of special requirements.

2.8.2 Cut Back Asphalts

The viscosity of asphalt can be lowered by dilution (cut back). A product easy to handle is thus obtained. At the same time strong heating of asphalts prior to its utilization becomes superfluous. After application the diluent evaporates and leaves the asphalt binder on the surface.

The dilution of asphalts with solvents and diluents is especially necessary in the production of asphalts for road purposes. Besides, cut backs are commonly used in making asphalt containing dyes and varnishes. Special units deal with the production of these. The general principles of making cut backs used in road construction shall be gone into here, since these materials are produced in the petroleum industry on a large scale and are delivered to the consumer. Cut backs are obtained from standard asphalts by diluting with suitable solvents. Specifications for cut backs relate to various groups on the basis of evaporation and bonding rates, respectively, as well as on viscosity values, considering the requirement for road construction. Recently the term "liquid asphalt" has been introduced for general use instead of the name "cut back asphalt". The latter term is still applied here with reference to the manufacturing process.

Different cut backs are made in the individual countries. Considering the evaporation of the solvent with various boiling ranges used to cut back the asphalt, slow-, medium-, or rapidcuring (evaporating and binding) asphalts are distinguished in the USA[14]. Slow- and medium-curing qualities are specified in the Soviet Union[15]. With a view to

TABLE 39.

Comparison of the Specification for Slow-curing Cut-back Asphalt (GOST 1972-52) With Experimental Results of Nagylengyel Crude and Residue

	GOST 1972-52 B-2	*Nagylengyel crude*	*GOST 1972-52 B-6*	*Nagylengyel residue (20°E at 100°C)*
Viscosity on STV (5 mm nozzle) at 60°C max., S	5–15	9.9	100–200	150
Distillation up to 360°C (Vol. %), max.	35	14.5	5	0
Residue Penetration at 25°C, 0.1 mm		liquid		
Floating test, min. S	20	53	50	50
Flash point, °C	70*	76**	120*	200**
Water soluble, max., %	0.30	0.08	0.30	0.10

* Brenken
** Marcusson

climatic conditions and road construction requirements, the above classification is altered and only contracted groups, or even a single group is used to design the satisfactory quality in other countries. Therefore only the slow-curing oil will be mentioned, called road oil in practice. The mild requirements specified for this quality render possible the direct use of a suitable crude, a crude oil derivative, or a distillate residue in some cases. Such a possibility is indicated e.g. in the Soviet standard specification literature(16). In the beginning of processing Nagylengyel crude oil with high asphalt content in Hungary, crude oil itself and its residue were specified as the so-called petroleum asphalt proper. Experimental characteristics based on the studies of Zakar(17) are shown in Table 39, together with specifications covering the group GOST B (slow-curing).

Some qualities meeting the standard specification may be produced by blending asphalt and crude oil depending on the crude oil quality.

Similar experiments carried out in a Baku plant are reported by Mihailov(18). Rapid-, and medium-curing qualities cannot be obtained by the process mentioned. Their manufacture is covered by the standard specifications of cut back asphalts. A paving asphalt with definite quality and a solvent of suitable boiling point appropriate for this purpose must be selected to obtain these brands. They are blended in a certain proportion to get a product with the desired boiling point(19).

TABLE 40.
Production of Medium-curing Cut Back Asphalts (ASTM D 593-46) From Nagylengyel Feed Stocks

Designation		*MC-1*	*MC-2*	*MC-3*	*MC-4*	*MC-5*
Asphalt		*Solvent*				
Penetration at 25°C, 0.1 mm	*Boiling point, °C*	*quantity, %*				
224	159–276	32+	26–27	20–22	17–18	3–15
	166–309	—	—	—	14–16	2–7
171	159–276	—	—	—	17–18	5–15
	166–309	—	—	20–21	15–21	5–15
113	159–276	—	—	—	—	5–15
	166–309	40+	27–32	24–29	19–24	7–17

In a far-reaching investigation series, Zakar and Mózes[20] established the production conditions of a medium-curing cut back asphalt corresponding to the ASTM specifications to be obtained from Nagylengyel asphalts and distillates.

Table 40 comprises data on three asphalts of various penetrations and on two solvents of various boiling points, the feed stocks for the production of the desired cut back asphalt, as well as the required quantities. The products correspond to the ASTM specifications for medium-curing cut back asphalts having different viscosities.

In a report on the manufacture of cut back asphalt, on the basis of data obtained in the plant described, Brooks[21] stated that to obtain rapid-curing qualities, a petroleum distillate with a boiling range from 102 to 219°C is required, whereas 174-302°C boiling range distillate is necessary when medium-curing material is desired. According to Traxler[22], 25-120 penetration asphalt has to be used for rapid-curing cut back asphalt, whereas 120-300 penetration asphalts are suitable in general to obtain medium-curing cut back asphalts.

The question of insuring the required solvent also arises, since the development of cut back asphalt production depends on the available solvents. Beside gasoline, kerosene, and gas oil, obtained in petroleum refining, application of tar oil distillates prevails. A favorable possibility to utilize crude oils with high asphalt contents consists in applying part of the distillates obtained immediately as cutters for the asphalts produced at the same time. Handling of asphalts in the plant and their dispatch is facilitated to a great extent by this practice. The manpower requirement is favorable in such an asphalt unit, since its automation can be carried out easily.

2.8.3 Asphalt Emulsions

Dispersing (emulsification) of asphalt in water is used to lower asphalt viscosity beside cutting with solvents. Asphalt emulsions demulsify more or less rapidly, in contact as a thin film with the surface of solids. This process is called emulsion breaking. The advantage of asphalt emulsions as compared to cut back asphalts consists in their non inflammability and their ease of coating wet mineral matter. Another advantage of asphalt emulsions is that they bond rapidly on the road surface and no heating is required. Thus the heating equipment connected with construction operations may be eliminated, neither must the aggregate be dried previously. Emulsions form a group in the disperse systems. Coarse disperions in which both the dispersed phase and the dispersing medium are fluid, are called emulsions. Dispersing liquids and the production of emulsions are called emulsification.

Emulsions having useful concentrations and stability can never be made without the application of some auxiliary material. These auxiliaries are generally called emulsifiers. They include all substances which apart from their emulsifying effect maintain the dispersion grade of the emulsion to render it stable in storage. This relates especially to the auxiliary substances commonly used in the production of asphalt emulsions and to those acting as emulsifiers and stabilizers at the same time(24).

The number of agents used for emulsifying has increased suddenly during the last decade owing to intensive research work in this field. The emulsifiers can be classified according to Becher(25) into the following large groups:

1) Surface active agents
2) Natural substances
3) Finely dispersed solids.

It must be mentioned that the above groups are rather arbitrary; quite a number of natural substances are surface active in some way. The surface active emulsifiers represent the type used in industry. Emulsifers can also be classifed according to the hydrophilic groups included in the molecule.

An emulsifying agent must have special solubility characteristics, that is, it must be compatible with both the oil and water phases.

They are characterized by having in their molecule a polar, water-soluble (hydrophilic) portion and a non-polar, oil-soluble (hydrophobic) portion. They are surface active in the sense that they have the ability to migrate or travel to a surface which is the interface between two liquids. In asphalt-water systems, the emulsifier moves to the interface between the asphalt and water. The hydrophobic part of the emulsifier molecule embeds itself in the asphalt surface imparting to the surface an ionic charge. The hydrophilic portion of the emulsifier extends out into the water phase[26].

Emulsified asphalts may be of either the anionic, electro-negatively charged asphalt globules, or cationic, electro-positively charged asphalt globules types, depending on the emulsifying agent. The non-ionic type is used on an extremely limited scale in asphalt emulsions.

An emulsified asphalt, in which the continuous phase is asphalt, usually all cut back (RC or MC) asphalts, and the discontinuous phase is minute globules of water in relatively small quantities, is called an inverted asphalt emulsion. They are water-in-oil type emulsion. This type emulsion may also be either anionic or cationic. In general, asphalt emulsions are of the oil-in-water type, that is, asphalt is the dispersed phase as globules suspended in an aqueous continuous phase[27].

2.8.3.1 Anionic Asphalt Emulsions

The general structure of the anionic (alkaline) emulsion type of emulsifying agent is represented by the formula

$$CH_3(CH_2)_{14}COONa$$

The simplest way of making anionic agents is to disperse saponifying acid containing asphalt in liquid alkali. The neutralisation process promotes the dispersion of asphalt in water. Of course, vigorous agitation is required during the operation. In this manner, soft asphalts with a natural acid value in excess of 0.8 are easily emulsified even in very simple units to give suitable emulsions. The emulsibility of an asphalt having an acid value from 0.5 to 0.8 can be augmented by adding small quantities (0.1%) of fatty acids.

Several commercial products are available for this purpose, such as oleic acid, sulfonates soluble in the oil obtained from acid tar, high molecular weight naphthenic acids. Although the emulsibility of asphalts is improved by these substances, their effect is smaller than that of the natural naphthenic acids of asphalt. They may even prevent emulsification when added in higher quantities such as 0.5%. The above mentioned agents are ineffective with asphalts having an acid value below 0.5, and higher molecular weight acids must be added in such cases. Considering the acid content of most asphalts, soap solutions containing free alkali, generally potassium hydroxide in excess, are used. The quantity may be 1-2%.

2.8.3.2 Cationic Asphalt Emulsions

In the last decade the application and breaking of road asphalt emulsions was the object of far-reaching investigations, leading to the production of new kinds of asphalt emulsions. The purpose of these studies was to develop good adhesion and a rapid bonding asphalt layer. Later it was considered a drawback that according to the electrokinetic theory anionic emulsions exhibit good breaking properties on basic agglomerates to form a hydrophobic asphalt film. However with acidic mineral matter they were not quite satisfactory.

The emulsion breaking tests on various agglomerates, the research in the field of acid phase emulsions, as well as the good results of adhesivity of cationic substances resulted in the development of a new emulsion type, the cationic asphalt emulsion[29].

As far back as 1932 the French investigators Vellinger and Flavigny expressed the desire for acidic asphalt emulsions. However, these were only realized when on the initiative of Duriez and the good experiences gained with fatty amines in England, the technical difficulties in the production of cationic asphalt emulsions were overcome to an extent, their application in practice could be started, especially in France. It has been increasing since 1952. Closely connected with the manufacture of cationic asphalt emulsions, the question of their application arose, and gave rise to considerable research work in this field[31,32,33].

Contrary to the alkaline reaction of the anionic emulsions used commonly in the beginning, the water phase of cationic asphalt emulsions is acidic. Corresponding to this, the limits of stability of cationic emulsions are in the range of 2-6 pH, as against those of the anionic emulsions which are between 8 and 12 pH.

The advantage of cationic emulsions consists in the fact that good adhesion results together with the breaking of the emulsion on nearly all aggregates even under circumstances of extreme humidity[24]. The earlier opinion considering only asphalts of high acid value suitable for emulsion production viewpoint had to be changed. This value is indeed very important when emulsifying in alkaline media, but is of no consequence if emulsifying takes place in an acidic phase. High acid value can even interfere under certain conditions. The general structure of the cationic (acid) emulsion type of emulsifying agent is represented by the formula

$$CH_3(CH_2)_{17}NH(CH_2)_3NH_2H$$

Mainly high molecular weight fatty amines of 12-18 carbon atoms are used as cationic emulsifiers in the production of cationic asphalt emulsions. Emulsifying agents are generally amine salts made by reacting hydrochloric acid or acetic acid with an organic amine or diamine. This is usually accomplished by mixing the asphalt amine mixture with dilute acid solution forming the amine salt emulsifier in situ.

These amines are also made from fatty acids, which are used as feed stocks in their preparation. Another kind of cationic emulsifier is represented by the quaternary ammonium salts.

2.8.3.3 Natural Substances and Finely Dispersed Materials

Many of the natural substances are relatively ineffective by themselves. Their joint application with other emulsifiers is however very favorable. These compounds are therefore often called auxiliary emulsifying agents. The viscosity of the emulsion often increases under the influence of natural substances and settling of the emulsion is retarded.

Sodium lignate is often added to alkaline emulsions to effect better emulsion stability. Cellulose derivatives are also used to increase the viscosity of the emulsion if needed. If very stable emulsions are wanted in application, the emulsifier will mostly be albumin such as casein, animal or plant albumin, and globulin or animal glue in an approximately 4% solution. In the case of sulfite liquor application, emulsification takes place in a weakly acidic medium. If solutions with sufficient concentrations are used, very stable dispersions will result. To avoid settling, a clay suspension is added in such cases.

It is to be mentioned here that apart from the previous classification of emulsifying agents there is another distinction based on their manner of action. Thus there are low and high molecular weight emulsifiers, which may be natural substances as well as synthetic products. The third group is based on dispersion characteristics. Quite a few finely dispersed solids are effective emulsion stabilizers. Very much harder asphalts can be used when insoluble emulsifiers such as clay are applied. Full particulars were given by Drukker[35] on clay emulsion preparation and properties. Commercial production of such asphalt emulsions has been expanded by refinement of the early methods in processing equipment, availability and treatment of mineral colloid clays, availability of many types of asphalt materials and a broader knowledge of emulsion performance and use. Clay-type asphalt emulsions are used as protective coatings, roof coatings, flexible binders for flooring mastics, and insulating compounds.

2.8.3.4 Asphalt Emulsion Production

The bulk of asphalt emulsions is produced mechanically by dispersing asphalt in water (diluted emulsifier) at a carefully regulated temperature. The asphalt is dispersed by means of a high efficiency emulsifier equipment, mostly a colloid mill, but even by a simple mixer or by a homogenizer in some cases. The colloid mill is the device most often used to manufacture asphalt emulsions. Although there are several types of colloid mills, they all work on the same basic principle, a high speed rotor moving within a fixed stator(36,37,38). The rotor turns at speeds of between 1000 and 6000 revolutions per minute. The clearance between the rotor and the stator is about 0.3-0.5 mm. The high shear breaks the asphalt up into fine droplets which are then readily stabilized by the emulsifier contained in the water phase. The product has a particle size of approximately 1 to 2 micron depending upon the equipment size and the conditions of emulsification. Acoustic energy(39) is also used to make asphalt emulsions up to a certain capacity. One of these types is the Econosonic industrial unit recommended for the production of both anionic and cationic asphalt emulsions. Here the only moving parts are in the simple gear pump used to power the Sonolator. The Sonolater itself has no moving parts to wear out. Acoustic energy as generated by the Sonolator is the most effective and efficient type of mixing energy in current use(40).

The selection of the type of equipment is dependent on the expected volume of manufacture as well as on plant investment considerations.

The capacity of pilot colloid mills generally amounts to 1 t/hr that of production mills to 10-15 t/hr. The economic limitations of watery emulsion transport must also be considered when determining the capacity.

Asphalt emulsions are nearly always manufactured in special plants and not in the asphalt unit proper. In a plant to produce asphalt emulsions, it is necessary in addition to the mill to provide tanks of sufficient capacity to hold the asphalt, emulsifier and the finished emulsion. It is advisable to provide for the asphalt supply by means of asphalt in a hot state. An extra tank of large capacity is made available for this purpose. It will be advantageous to have also a small tank with

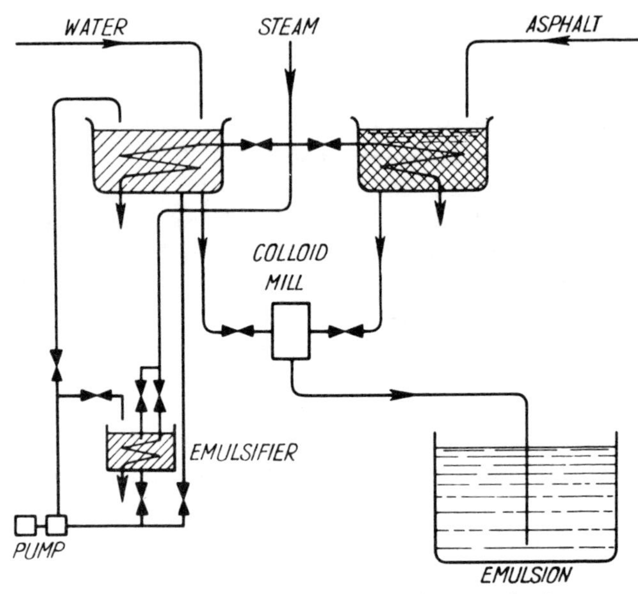

Fig. 37. Flow sheet of cationic emulsion production

recycling pump and steam heating open and closed coils to make the emulsifier. As a rule it is preferable to have the emulsifier and asphalt tanks overhead and the emulsion tank on the same level as the mill so as to avoid a fall in delivery due to a too great delivery pressure[41]. The flow sheet of cationic asphalt emulsion production is given in Figure 37. The dosage of asphalt and water containing the emulsifiers is either made by separate pumps, or these pumps are fixed on the colloid mill itself. As soon as asphalt flow is shut down, the mill is allowed to run a few minutes on the diluted emulsifier to wash out the mill and lines and then the wash liquid is circulated to a waste tank.

The same equipment may be used in principle to make both anionic and cationic asphalt emulsions. If cationic emulsion production has to be taken up after the manufacture of anionic emulsions, it is very important to clean these units carefully before switching over, since mixing the two kinds of emulsions might break the emulsion produced. All parts of the equipment used previously to make anionic emulsions, such as tanks, lines, valves, and the colloid mill must be washed thoroughly with hot water and treated with dilute hydrochloric acid to neutralize any material that may remain in the equipment[42].

The production unit must provide for a certain storage of the emul-

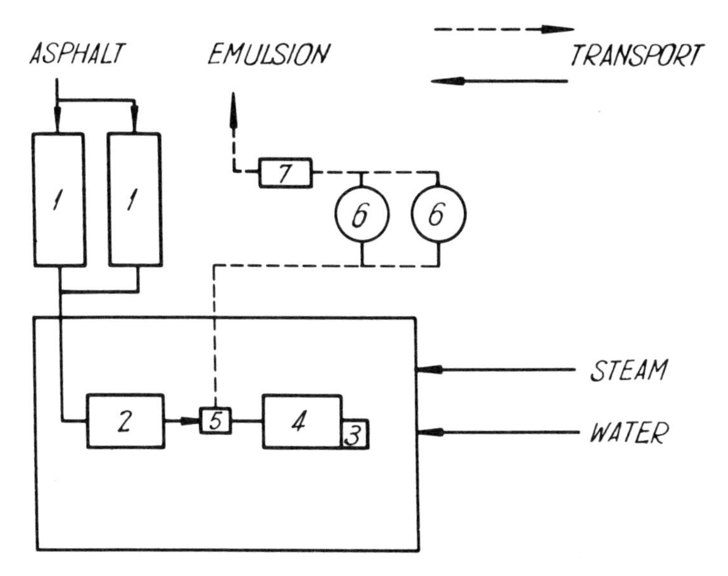

Fig. 38. Layout of asphalt emulsion plant
1. Asphalt storage 2. Asphalt-phase tank 3. Emulsifier 4. Water-phase tank 5. Colloid mill 6. Emulsion product tank 7. Pump

sion. The extent of the storage depends in the first place on the cooperation between production, transport, and utilization, as well as on probable changes in weather.

The layout of asphalt emulsion plants can be seen in Figures 38 and 39, considering the most important units. Depending on the local circumstances, similar or different allocations can be realized. Outside steam energy is used in the layout of Figure 38, while the unit given in Figure 39 has its own steam unit. The common feature of the two allocations is the central production part in a protected position, where the preparatory operations which are not shown separately on the layouts, emulsifying and dosage of acid, can also be carried out. A common solution for the ingoing transport of hot asphalt and the outgoing transport of finished asphalt emulsion is represented by a single road used. In the course of asphalt emulsion storage, certain phenomena may occur, which must be taken into account in storage room design and in the connected manipulations. The asphalt emulsions made can generally stand a storage of three months in practice. Special measures should be taken only if the emulsion cannot probably

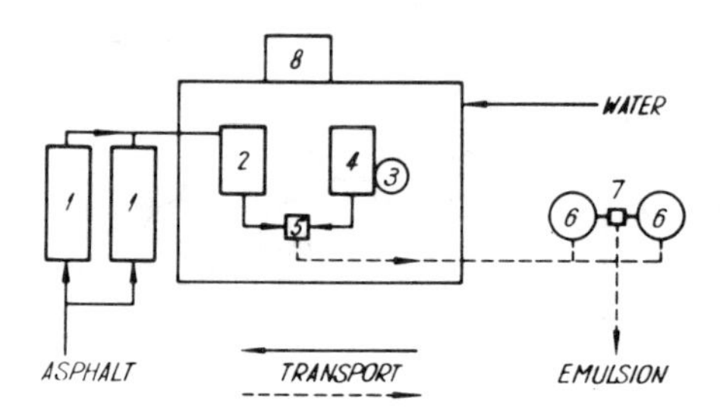

Fig. 39. Layout of asphalt emulsion plant
1. Asphalt storage 2. Asphalt-phase tank 3. Emulsifier 4. Water-phase tank 5. Colloid mill 6. Emulsion product tank 7. Pump 8. Steam-centrum

be stored during the usual period. Loading and transport must be brought about in due time in this case.

Although the storage is in accordance with the specifications, the formation of a thin surface film and of a layer with higher water content in the upper part due to the separation in these layers has to be considered. This, however, is no obstacle in itself, and it needs not to be broken if not required specially. In some cases a film of heavy kerosine is placed on the surface of the emulsion as a seal.

Vertical tanks are thus used for the storage of emulsions, and a pipe reaching the bottom of the tank is used for feeding the material[42]. Foaming must be avoided in the manipulations by careful handling. Settling can be eliminated by stirring in emulsions suitably made. Its extent depends on storage time. Various means may be applied for mixing. Rotational positive displacement principle pumps are preferable in pumping.

Slight heating of the material contained in the tank must usually be done. Undesirable cooling of the stored emulsion may result in breaking. The material therefore must not be allowed to cool below 5°C. Heating by hot water is adequate, but steam heating is also used. In the latter case, care must be taken to avoid overheating and local heat accumulation, since this could cause emulsion breaking to a certain extent. Apart from these methods, electric heating is also used in some plants.

Heating possibilities must be provided considering asphalt precipitations and plugging. Periodical cleaning of emulsion tanks, of lines, and of pumps are an absolute necessity for keeping the equipment in perfect repair.

Almost all grades of asphalt can be emulsified from the hard penetration grades to soft grades of straight run asphalt and the very soft cut-back-asphalts. Harder grades of asphalt are more difficult to emulsify, and the emulsions made from them are generally used for industrial purposes[44]. Emulsions made from asphalts of 180-220 penetration are normally used in road making. The penetration of the asphalt used may differ from this range in both direction depending on the climate and on road making conditions in this connection.

The higher the asphalt penetration, the easier it is for the asphalt to coat aggregates. However, strength requirements of the finished work needed to sustain heavy traffic demand a relatively low penetration asphalt to give high asphalt cohesion strength in the rolled surface. These two requirements in the penetration grade of the asphalt are divergent, but this can be simply overcome.

A low penetration asphalt can be softened to improve aggregate coating and mixing stability by the inclusion of small amounts of naphtha or kerosene cut back in the emulsion formulation. Knowing the service requirements, the determination of the viscosity of the asphalt used is very important. Naturally further differences are experienced apart from the penetration group in the case of asphalts of various origins. In order to achieve adequate emulsification, the setting of the desired viscosity is of outstanding interest apart from the suitable device. This involves in turn the knowledge of the temperature required, and its maintenance during production. According to a rule of thumb, the sum of asphalt and emulsifying water temperatures should not exceed 200°C. The water solution temperature must be kept at a lower value in the case of asphalts of higher viscosities. It has been mentioned that the emulsion making formula must be fixed corresponding to the emulsion type to be produced and to service conditions.

Considering the various asphalt types utilized and the great number of possible emulsifiers, the previous tests should always be carried out. The aim of these is to render possible the choice of the most economic production conditions for the desired quality, including the type and quantity of the emulsifier required.

Together with the individual emulsifier types, the emulsifier quantities used for the individual emulsion types are also recommended by the emulsion manufacturer[45,46,47] Considerable research work is going on to clarify the interconnected factors in asphalt emulsion making. Results obtained so far in the research series covering the correlations of production conditions are published by Sauterey and co-workers[48,49]. In the first place results on cationic asphalt emulsions have been given by the French researchers.

Most commercial asphalt emulsions contain 60 to 65% asphalt. Independent of this, asphalt emulsions with 50-75% asphalt contents are also made to meet the requirements of the users[50]. The making of emulsions having higher asphalt contents was facilitated by the introduction of cationic emulsions. The viscosity of cationic emulsions is lower than that of anionic emulsions of similar asphalt content. This means that by the use of cationic emulsifiers it is easier to obtain emulsions of 65%, or higher asphalt content, without the emulsions being too thick to handle.

Considering that asphalt emulsions are generally made in plants built for this purpose, problems in connection with the emulsions will be taken up here, apart from the separate treatment of filling and transport of other asphalt types.

Emulsion transport may be effected in railway tankers, in tank cars, and drums. Earlier drum delivery was most commonly used, nowadays transport by tank cars is more widespread. The 200 pound drums are common in drum transport, but certain special emulsions are filled and transported in smaller capacities. When the emulsion is packed in drums, aluminum bungs should not be used since they are attacked by the free acid, steel or plastic bungs are suitable.

Tank cars of various capacities are used for emulsion transport. In the case of large tank cars, the separation of the tank by inside partitions becomes unavoidable. This decreases shaking, and adverse phenomena may thus be avoided such as foaming and partial breaking.

Only such transport vessels and tanks are allowed for the transport both by drums and in tank cars as have previously been cleaned and washed thoroughly. Special care must be taken when anionic and cationic emulsions are produced and transported in the same unit.

Some investigators report on the manufacture of emulsions in their countries. It was reported by Lüder and Heerwig[51] that in East Germany anionic emulsions are made almost exclusively, chiefly the unstable types used in the paving industry. The asphalt emulsion manufacture developed in Kasakhstan is described by Agapov[52].

At present the Kasakhstan plants are able to produce 100,000 tons of emulsions in a building season. Their capacity can be increased by 50%, if the necessity arises. A more detailed description of the operation of emulsion making units is given by Raikov[53] as for a plant to produce emulsions for the construction industry. Information on a plant in the Volga-Don industrial area is published by Kejman[54] dealing with asphalt emulsion manufacture with surface active emulsifiers. Beside the Hotunov-Puskin dispergator, acoustic energy is also used. The emulsion quality is improved by the latter due to more accurate dosage.

2.9 Asphalt Service

Beside the production plant proper, the service units rendering possible the operations are of outstanding importance. Storage, blending, loading, packaging, and transport of the asphalt made is effected by means of these. Another unit has to be built in most cases to make the steel drums necessary for charging asphalt. Depending on the manner of packaging and transport, respectively, these service units may represent the largest part of the plant. It is clear that the design and operation of these plant parts is an important factor for the adequate layout and operation of an asphalt plant.

2.9.1 Tank Farm

The tank farm joined to the production units proper is an important part of the asphalt unit just as it is in any petroleum refinery. One part of the tanks serves the purpose of intermediate storing and ensuring the necessary blending components, and another part is used for the storage of finished products. The arrangement of the intermediary and blending tank park, and the size of the tanks used depends on local conditions, the special asphalt qualities to be made, as well as on the demand for the produced qualities. The tanks utilized for road asphalt are larger, and those for blown asphalt are smaller. The blown asphalt is not charged into intermediary storage tanks in some units, but drained directly from the vessel after being cooled to a given temperature.

1000-5000 m^3 tanks proved satisfactory for storing and blending low viscosity road asphalts, and 100-500 m^3 tank capacities seem to be best for high viscosity qualities. Tanks of 80-250 m^3 are used to store and blend blown asphalts[1]. Smaller tanks are equipped partly with internal mechanical mixers. Many tanks are elevated to permit gravity loading and have slanted bottoms to facilitate evacuation. The size of the tank farms varies between 5000-20,000 m^3 [2]. The farm size depends on the production volume, the demand, and the fluctuations according to seasons. Corresponding to the seasonal character of asphalt demand, the largest storage capacity must be available during the summer, whereas less capacity is required in winter[3].

Asphalt tanks have to be heated, independent of their being insulated or not. The tanks are heated by various methods. A central heating equipment may be used, if this is rendered possible by the location of the field storage tanks. Mostly a capacity of 1.5 Mill. kcal will be suitable, with 2.5 Mill. kcal in the case of higher requirements. If the tanks cannot be arranged centrally, separate heating for each will be used. A simple internal flue type consisting of a horizontal firing tube extending across the tank bottom and rising through the tank roof, which forms a stack, may be selected. Asphalt must never get below the firing tube level during heating. Individual tank heaters with outside firing are also used. Heat is supplied by them to a shell through which the asphalt is circulating. The tank is kept warm by electric heating in some units, or it is heated in this manner when the necessity arises.

If enough steam is available, various heat exchangers are inserted into the tanks. The heat quantities necessary for keeping the asphalt in the tank in a hot state are thus supplied, or heating the asphalt is rendered possible in this manner. Bundle heaters proved suitable for this purpose. Asphalt is pumped through the bundles and thus heated to the necessary temperature for further processing. Hot oil systems are used successfully in certain units. Oil is heated to about 300°C in a vertical flow heater and supplied to the individual points by circulation pumps for use. A temperature of approximately 200°C can be maintained in the tank. An expansion chamber-floating on the return line just ahead of the circulating pumps-absorbs changes in volume of oil in circulation as usages change. The contact between the oil and the air is kept at a minimum during circulation. The oil surplus flows from the expansion chamber into a storage tank. If loss occurs, this can be made up automatically from the storage tank[4] Whether heating is effected from outside or internally, or arranged centrally, steam heating is considered the best way of heating from the cost viewpoint. The waste heat of tube stills in the refinery distillation units is applied by Ney and co-workers[5] to heat the oil for the hot oil heating system of asphalt tanks. During off time, heating the oil takes place by a vertical tube still. Hot oil heating is used for another reason; swamping of the asphalt and foaming as well as the explosion danger are eliminated.

The tanks are usually insulated. This is absolutely necessary with tanks which are often unloaded. If otherwise, the asphalt proper solidified on the tank walls supplies a suitable insulating layer. The insertion of mixing devices or the arrangement of circulation systems is advisable for the purpose of reaching uniform temperature distribution and uniform quality of the stored asphalt.

It should be mentioned here that the discharge or circulation lines must not join the steam zone of the storage tanks, as is known from petroleum refining. Should feeding near to the liquid level be desirable, adjustable inlet tubes can be used. Hazardous situations may frequently arise with asphalt tanks under service conditions, thus blowing off often occurs. In less dangerous cases the excess pressure can still be expanded through the vent. The resulting suction effect is often too strong for the vents and the tank roof is drawn in. No fire can occur due to air shortage if the roof does not break.

With heavy blowing off, the roof is usually partly broken. In such cases the asphalt on the liquid surface catches fire, but the fire can be put out easily with a CO_2 extinguisher or with steam. Application of foam and even water involves the danger of foaming and is thus very hazardous.

The reason for these blowings is not completely clear. A material of 250°C or of even higher temperature has often been pumped into the tank and mists with wide inflammability limits may be formed there. Iron sulfide is often considered to give rise to conflagrations. According to other information, it could be proved on the occasion of such an accident that water in the tank was suddenly heated to above 100°C by the hot product and rapid evaporation was the result. The tank broke due to the pressure increase. The burning itself may have been caused by a spark emitted by bursting. A defect of the heating coil proved to be the primary reason for the fire as a large enough water quantity entered the tank[6].

Great care must be taken in the design and maintenance of pumps and tubes also. Pumps used in plants are mostly piston pumps driven by steam or electricity. Gear pumps are also often used. Rotary or reciprocating, positive-displacement type pumps of various constructions appear to be rather advantageous. Pulsation-free flow is best provided for by this pump type. This is of great importance especially in the case of automatic mixing units. The pumps used in plants are supplied with heating jackets on the product side and at the vents.

Product feed lines must be above ground, since the maintenance of lines laid underground is difficult and expensive. The lines have to be heated and insulated. Where high viscosity low penetration asphalts are to be transported, insulated lines with steam jackets are absolutely necessary. Drawbacks to the steam jackets are the high investment costs and difficulties in maintenance. Conduits supplied with internal heating are also employed, but they do not seem advisable because of damage risks and maintenance. Trouble may arise in the case of conduits with steam jackets or internal heating due to steam or condensate that enters the product inevitably from time to time. To avoid this danger, so-called accessory heating is used in which one or more heating conduits made of copper or steel are mounted at the product line and are insulated together with the latter. Recently heat transfer cements are recommended. Heat transfer cements have the property of conducting heat to the surface to be heated. The purpose is to efficiently transfer heat directly into a pipe wall by conduction rather than by convection or radiation.

Beside these heating possibilities, flexible electrical heating ribbons are also applied for the heating of definite sections in asphalt conduits.

2.9.2 Blending and Mixing Units

The tank farm is joined by the blending and mixing unit, the purpose of which is the homogenization of the produced asphalt, the blending of asphalts of various origins and penetrations, and blending the asphalts with petroleum distillates. The size and location of the tank farm of the blending unit belonging to the asphalt plant depends on the nature of the blending unit. Beside blending in tanks, the continuous automatic in-line blending system spreads to an ever increasing extent in asphalt plants. In general, the blending unit should be very near the tank farm. This is advisable from the standpoint of investment, operation, and control.

2.9.2.1 Blending in Tank

In the production of road asphalt qualities, often 200-300 penetration and 40-50 penetration asphalts are obtained by direct distillation and then stored. The desired quality is made by blending these two asphalt types before transport[7]. In the conventional method, the blending tank is filled with the products stored in the component tanks at a definite proportion. Exact composition of the blends is effected by a suitable flow measurement device. An electrically driven mixing apparatus is operated together with the flow measurement device during the whole period of filling. Upon completing the loading, a sample may be immediately taken from the tank to effect a control test in the laboratory. Mixing in tank may be done by a circulation pump or by an injection mixer inserted internally. However, mixing by air has also good results if used in a suitable manner. The mixing time must however be relatively short, since the properties of the asphalt are changed on contact with air. Again, care must be taken that no breakdowns should be caused by the condensate deposited in the air conduits. Therefore suitable knockout drums, drying equipment or other means must be provided for to keep entrained moisture controlled. Cut back asphalts are produced by blending the base asphalt with a diluent. Asphalt and diluent are often introduced separately into the blending tank. In such cases asphalt is often added to the diluent to make use of the cooling effect of the cutter and to decrease solvent losses.

Blending must always be made rapidly. Beside electrically driven devices, mechanical, or internal jet type mixers, mixing by air is often used also. Blending must be carried out quickly also, since especially with rapid curing oils, solvent losses can be very considerable. Long storage of cut back asphalt is to be avoided if possible, or else the material must be homogenized again at a later date, and the quality of the material is also changed due to solvent losses.

2.9.2.2 Continuous Automatic Blending

The continuous blending methods applied commonly in the case of fuels and lubricating oils can be utilized also for blending asphalt. Suitable storage tanks have to be allocated to provide automatic operation. Two tanks are required for each feed stock. One of them is filled with the material and the latter is pumped after an analytical test for blending. Then it is discharged from the other pump. The feed stock in the tank must be stirred in a continuous manner especially with large tanks, to maintain uniform quality of the material tested. The temperature of the material must also be carefully maintained constant. The blending unit proper ought to be built near the loading. The conduits and armature systems connecting the tanks must be insulated and heated suitably. The feed stock is pumped in the recirculation conduit by a pump between the blending section and the tank. The cutter is also kept recycling. The meter controlled components fed in a given ratio are blended in the mixing section. Upon homogenization, the mixture is discharged into the tank cars from which control samples must be taken prior to loading.

It the blending proportion is interfered with by some trouble, the

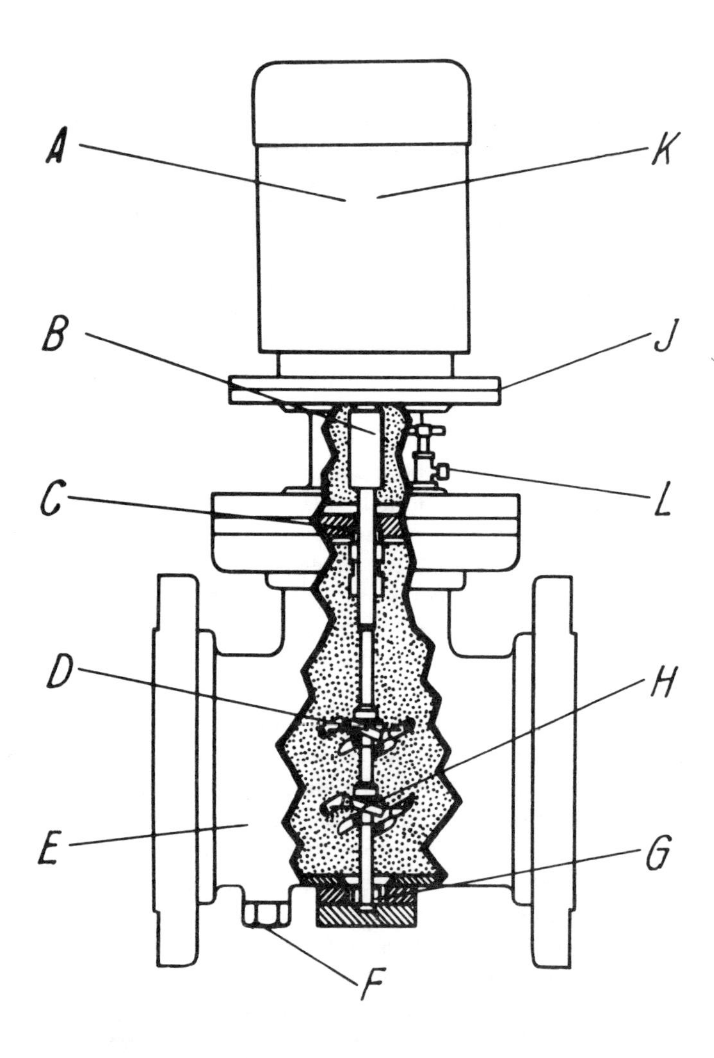

Fig. 40. In line mixer (Plenty)
A. Drip proof, totally enclosed, weatherproof or flameproof motors B. Sleeve coupling C. Mechanical seal for maintenance free services, or stuffing box and gland D. Polished stainless steel "Impelator" heads on shaft E. Body can be jacketed for heating F. Drain plug G. Shaft runs in steady bearing H. Two or more "Impelator" heads J. All flanges spigot mounted for foolproof replacement K. Air motors for flameproof areas L. Air release valve

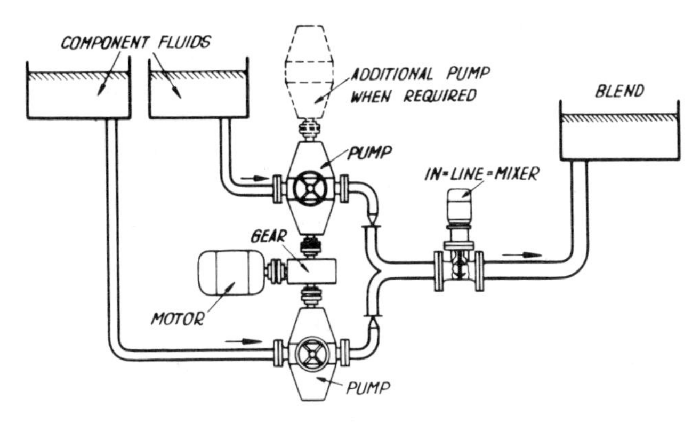

Fig. 41. Layout for continous blending of various asphalt feeds

mixing unit will shift to recirculation automatically. When the total desired quantity has passed, the blending unit is also stopped automatically[8,9,10,11,12].

A Plenty in-line mixer[13] is illustrated in Figure 40. It is used installed direct into the pipe lines and provides thorough and homogeneous mixing. The in line mixer contains the mixing head Plenty, which produces centrifugal force, suction and shearing action in a countercurrent whirl, and results in very rapid and thorough mixing. This method is used primarily with cut back asphalt blending but may be used also for mixing of paving asphalt components.

The layout for continuous blending of various asphalt feeds is illustrated in Figure 41[14]. The system comprises two variable capacity Plenty pumps all driven at the same speed by a common motor. Each pump draws a particular component fluid from its respective tank, and delivers at a measured rate into the common delivery line, through a non-return valve. Immediately the fluids flow together, they enter an in-line mixer where they are rapidly mixed before being delivered to the blend tank.

A great advantage of this system consists in the fact that the manufacturers need not provide for the storage of various finished goods. By starting up and suitable controlling of the blending and loading system, any demand may be satisfied at once[15]. The blending unit and the loading cannot meet the requirements without a laboratory working rapidly and accurately to enable continuous control of the raw material and finished goods. Special care must be taken with its installation and its equipment.

2.9.3 Filling, Packaging, Transportation

One of the most important requirements associated with the commercial production and marketing of asphalt is the transportation of the asphalt made in the plant to the consumer. Asphalt is transported in bulk, in a hot fluid state in rail tank cars, by road vehicles and truck tank trailers or in a cooled, solid and plastic state packaged, or without packaging. The packaging methods used as well as the transportation of asphalt are shown in Figure 42.

The most advantageous method of handling asphalt is in the hot, liquid state. Thus not only the difficulties in the refinery may be avoided and great savings in packaging material realized, but the customers may recover the heat content of the asphalt transported in a hot state. Handling in a hot fluid state is effected whenever possible with straight run reduced asphalts, with blown asphalts of 75/30, 85/25, and 85/40 types, as well as in warm fluid state with liquid asphalts. This method cannot be resorted to in the case of high softening point asphalts and of certain special qualities; besides not all consumers possess handling

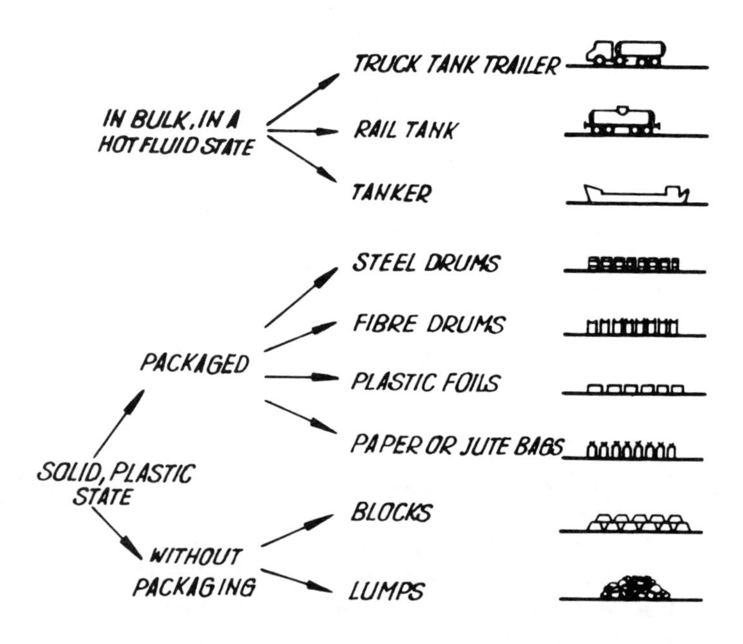

Fig. 42. Various packaging and transport methods of asphalt

TABLE 41.
Transportation Methods of Asphalt, %

	1953	*1958*	*1962*	*1964*
Truck tank trailers	1	43	79	85
Rail tank cars	63	40	13	9
Packaged asphalt	36	17	6	4
Tanker	—	—	2	2

facilities for hot fluid asphalt. Thus the plants must also provide for asphalt packaging(16) considering the various asphalt qualities.

The changes taken place in the last 15 years in asphalt transportation are illustrated by Table 41 giving data on the transporation of asphalts made in West Germany(17).

2.9.3.1 Bulk Transport in the Hot State

Special rail tank cars are required for the transportation of hot

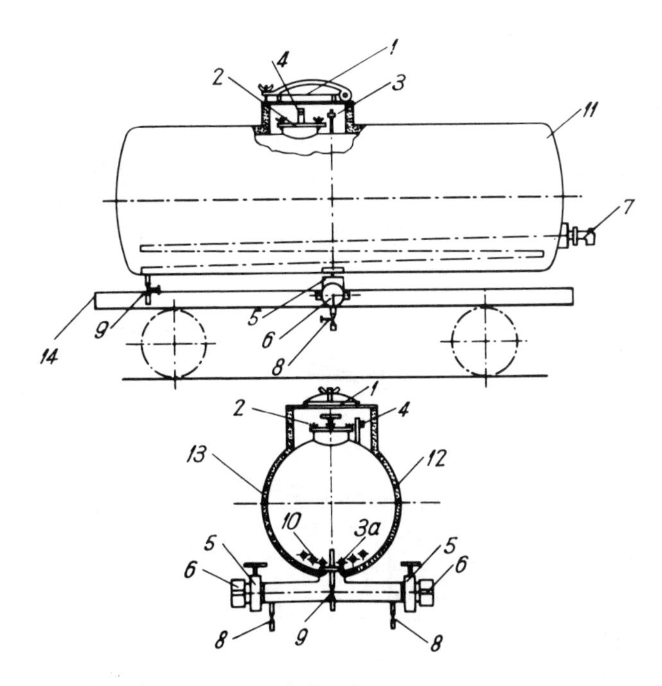

Fig. 43. Rail tank car with steam heating
1. Dome cover 2. Internal cover 3. Valve rod handle 3a. Internal valve 4. Safety valve 5. Outlet valve 6. Valve cap 7. Steam coils 11. Steelplate-tank 12. Insulation 13. Chassis

asphalt by railway. These special tank cars for asphalt must be equipped with a suitable heat insulation, a heating unit, and heated armature. Older cars have a capacity of from 15 to 22 m^3, but nowadays rail tank cars having 30-40 m^3 capacities are commonly used, and even rail tank cars with 63 m^3 capacity have recently been introduced[18]. The tank in the car is usually surrounded by a 100-110 mm thick insulation layer. Steam heating is generally used to heat the asphalt in the tank. Depending on the car size, a heating coil system with a surface of 10-35 m^2 is fixed at the tank bottom. The unloading line connected with the tank body, as well as the fittings can also be heated by steam. Beside steam heating, other heating systems are used with certain special tank cars, for example, direct flue firing heating. The arrangement of a rail tank car with steam heating for asphalt transport is illustrated in Figure 43.

Daily heat losses of the rail tank cars depend on the size of the cars, the quality of heat insulation, as well as on the initial temperature. They vary between 15 and 20°C. To facilitate loading of asphalt from the rail tank car, this can be pressurized to maximum air pressure of 1.50 atmospheres. The capacity of truck tank trailers used for the transportation of hot asphalt has reached 22,000 to 24,000 1b. These trucks are mostly trailer trucks, in which the motor section and the asphalt tank section are separated according to this construction[19] (see Figure 44). The tank body is surrounded by a 80-120 mm thick

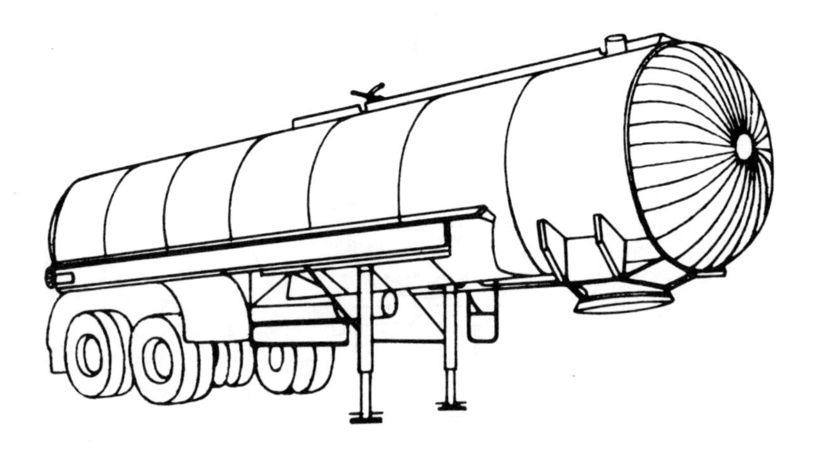

Fig. 44. Truck tank trailer

insulation layer, coated with protective sheet iron. At an ambient temperature of −20°C, a cooling rate of 1-1.5°C/hr may be expected. The trucks are generally provided for the intake of asphalt having a temperature of 200°C. In some cases the internal part of the tank is separated into two parts, having an opening that can be locked. This renders possible unloading the transported asphalt in two portions. The rail tanks are equipped with a compressor for 0.5 and 1.5 atm pressures. In some cases there are steam heating coils in tanks. Another solution is a heating flue tube along the middle of the tank. Asphalt can be transported in tankers built for this purpose both at sea and by inland navigation[20]. An example of the construction of such ships, a tanker 167 m long and 21 m wide of Swedish make with a capacity of 20.000 m^3 shall be mentioned here which serves the purpose of transporting asphalt from India to European and African ports[21]. A heating coil system is placed in the ten tank sections of 2,000 m^3 each. The temperature of the material can be maintained above 159°C by means of steam heating. The tanker is equipped with pumps of 250 t/hr and 500 t/hr performance, respectively, to handle asphalt.

The asphalt tanker from the series "Bitumina", built by the Bayerische Schiffsbaugesellschaft (West German), and operated by Shell, is 80 m long and 9.5 m wide. It has a capacity of 1200 t asphalt, which can be kept at a temperature of 180°C by an oil circulation heating. These tankers are equipped with pumps with a performance of 175 m^3/hr and have a travelling speed of 18 km/hr[22].

As it has been mentioned, the transport of hot asphalt requires the most simple facilities, since only a charging unit is required to load the rail tank cars and truck tank trailers suitable for transport. The loading temperature of the tank cars is such that it can be unloaded by the customers without further difficulties. This temperature must be ensured by setting the tank temperature at the desired value, or in special cases extra heating equipment must be available ahead of the loading point. Various hot oil systems proved suitable for this purpose[23]. Asphalt is loaded into the truck tank from the heated tanks by a pump or by gravity if placed at a higher level. In some units a truck tank trailer may be filled through the loading tube from the asphalt tank elevated above it within five minutes[24]. The rail tank cars are loaded directly close by the blowing tower in other plants, which provides for short loading lines. The various sizes of the truck trailer are to be considered when setting loading rates. The loading rate usually amounts to 800-1200 1/min. Higher rates require special measures to be taken.

The tank car loading place is usually connected closely to the in line system of the plant, especially if cut back asphalt is also produced. Loading facilities are arranged in an asphalt manufacturing plant erected recently[25] in the following manner. Six vehicle loading bays are provided. Two of the bays are for loading distillate, two for straight run asphalts which can be pumped from storage via a heater if necessary, and the remaining two bays are capable of handling either straight run asphalts or blended products.

The loading system is the facility for in-line blending. Two identical systems are provided each designed to blend four base components to produce intermediate grade asphalts and cut backs. Each blended has a maximum throughput of 2300 1/min. The system is based on the operating principle of a continuous comparison of pulse signals generated from positive displacement meters to a master demand signal.

2.9.3.2 *In the Cold State*

Special packaging media are generally necessary for the transportation of cold asphalt. In the first place standardized steel drums are loaded with a sheet thickness of 0.5 to 0.7 mm and a capacity of 200 1 all over the world. In some countries also closed and open fiber drums are used, respectively. The steel sheet or fiber drums to supply asphalt units are delivered from plants situated near the former but in numerous plants separate units are erected adjacent to the loading room, supplying the plant with the necessary drums.

The yearly production of a drum unit erected within the frame of an asphalt plant in Hungary[26] amounted to 500,000 drums in two shifts of 8 hours operation each, which enable the loading of 100,000 tons of asphalt per year. The diameter of the drums is 590 mm, their height 780 mm, and the thickness of the utilized steel sheet varzes from 0.5 to 0.9 mm. They are made from 1900 mm $\times$ 1500 mm $\times$ 0,75 mm iron steel sheets. Characteristic data for these sheets are

Tensile strength	36 kg/m^2
Stretching	34%
Erichsen Number	8.5

To lower packaging costs, the mass of the utilized drums is decreased to a minimum compatable with the strength requirements. Usually drums of 8.5 kg are used in the case of asphalt, but 9.5 kg drums of greater stability are advisable with cut back asphalts.

A firm equipment is made to load the asphalt to be marketed in drums, and the drums to be loaded are placed under it and removed subsequently.

The drums are forwarded by cart in older asphalt units, where the special type hand truck or wagonette has proved suitable, rendering possible the transport of one drum. The steel drums obtained in the transport of hot asphalt can also be used favorably as steel long range booms. In a number of plants the steel drums already loaded are allowed to cool and when the asphalt shrinks, the drum will be driven by the trolley back to the loading spots and refilled. Beside these truck or wagonette trolleys other types of hand- or motor driven trolleys are also used which enable the transport of four to six drums at the same time. In such cases several drums can be loaded together. Applying a series of such trolleys may result in a fair continuity of loading operations[27].

A roller conveyor is used for transportation purposes in the loading room in larger and more modern asphalt plants. The loading of asphalt takes place in a covered shed where the arriving empty drums are placed on the roller, delivered to the loading points and upon loading transported farther.[28]. Thus the asphalt drum runs in a continuous manner from loading to transport. Besides, electric carriages equipped with forked or plated levers, or lift trucks are used.

On loading fibre drums these are prepared and arranged round the loading points equipped with steel long range booms to be loaded by groups. Upon cooling of asphalt, the drums are transported and the loading points occupied by empty drums again. This method however involves a sufficient number of filling points being available.

A trend is experienced everywhere to substitute the expensive steel and fibre drums, by cheaper packaging material. Thus jute or paper bags impregnated with calcium chloride solution are used, which can readily be removed from the asphalt by the consumer[29]. The .inner surface of paper bags has been specially treated to facilitate stripping from the asphalt. Paper bags are used to handle from 25 to 30 kg asphalt. A special supporting structure is employed with the paper bags during filling. A definite temperature must of course be maintained in filling, since the hot asphalt filled at an excessive temperatures may burn the paper bags.

Upon development of plastic packaging materials, the asphalt industry resorted to packaging asphalt qualities into plastic foils. Usually, the polypropylene foil was placed in a suitable mold according to the earlier techniques. The asphalt was poured into the prepared mold, and it then assumed the shape of the mold. The ends of the foil were then folded and the piece packaged in this manner could be transported.

Thomas and Wilson[30] report on the development in this field. Preformed polyamide bags are utilized (Ultramid ex BASF) in the new method. The bags are filled with asphalt at about 170°C. The bags sink more and more during loading into the cooling water promoting heat dissipation. Upon loading the free ends of the plastic bag are cemented thermally and the packaged asphalt leaves the water bath gradually after complete solidification. The Ultramid prepared with various additives to suit the requirements is melted at about 220°C. This proved to be more advantageous than polypropylene melting in the range of 160-180°C.

One of the technical problems arising in connection with filling plastic foils is to handle and load asphalt at as low a temperature as possible. Ney and co-workers[5] report on a special continuous rapid cooling equipment for filling into foils. The cooled plastic asphalt is filled into the foil by means of a device. This packaging method insures the mechanization of asphalt filling and handling techniques to a great extent. Handling of 30 kg pieces is carried out by trucks with pallets.

Steel drums are suitable for filling and transporting all kinds of asphalt qualities. Again, fibre drums are only good in the case of various reduced asphalts and blown asphalts, excepting cut back asphalt. Paper bags and foil are commonly used for filling & transport of reduced and blown asphalts with higher softening points. In the warm season, the viscosity of otherwise plastic asphalts decreases; they become easily deformed and even fluid. During packaging and transport, this fact must be considered to avoid possible troubles caused by it. Higher softening point asphalts may be transported also without packaging. This can be done in various ways. The asphalt is filled in concrete pits, steel containers, or concrete basins, or perhaps in ducts (channels), or into lakes, where it will solidify. The asphalt blocks formed upon cooling can be lifted from the concrete molds. When pouring in ducts, the solidified asphalt must be picked and chopped up. Hard grades are run into lakes from which they are broken up into lumps for transportation loose[31].

The use of steel containers or molds proved to be the most advantageous. These are painted with chalk or whitewash, dried, and immediately filled with asphalt. Upon cooling, the asphalt is removed. The blocks can be stored directly by placing them on one another at suitable temperatures or placing boards between the individual layers.

The high vacuum asphalts and blown asphalts, as well as the qualities with 25, or 15 penetration and 75/30 in the cold season are stored in this manner. To avoid sticking together of asphalt blocks, various foils are used now for wrapping up the cooled asphalt blocks. If the asphalt wrapped in foils is melted at the customer's, no substantial quality changes are experienced, once the packaging material has been chosen well. It must be considered that asphalt blocks undergo some deformation in the case of very high ambient temperatures and long storage. To avoid rubble formation and attrition, both the loading and unloading of blocked goods must be carried out with the greatest care. Thus block asphalt should not be knocked over. The 100 kg size is often produced but mainly 30 to 35 kg blocks are asked for at the market.

The above mentioned finishing methods require a lot of work and more time as well as manpower than does loading, packaging and transport of asphalt in large quantities. The trend is manifest to automatise the processes as far as this is possible. Such methods could be introduced up to now only with certain asphalts of high softening points and low penetration. Some of these methods are similar to the pitch fragmentation used in the coal industry, in which hot pitch is allowed to flow in a thin layer on water cooled metal bands, where it is cooled. Crushing may follow. Another procedure is used in the pitch industry using extrusion of hot pitch through a thin opening into the cooling water with subsequent crushing. Since pitch and asphalt differ as to their properties, the process used with asphalts is somewhat different. The following method has proved satisfactory in the Hungarian industry. Hot asphalt is allowed to flow into water streaming in concrete pits. Then the asphalt band which floats on the water and has been cooled in the meantime is taken out by means of a suitable device,

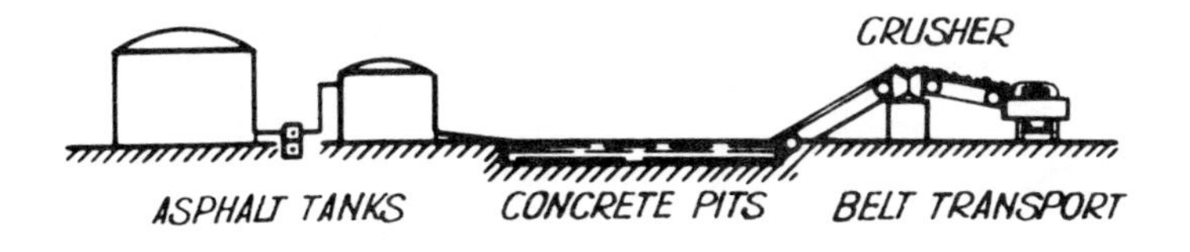

Fig. 45. The operation principle of an asphalt flotation unit

crushed, and forwarded in transportation cars[32]. Crushed asphalt obtained in this manner usually contains about 10% water[33]. The operation principle of an asphalt flotation unit is shown on Figure 45.

A method to produce commercial hard asphalt is to pass asphalt through a nozzle pipe, similar to the production of pitch in the tar industry, with subsequent cooling and spraying by air, gas, or steam to form the so-called flotation asphalt[34,35]. The packaging medium necessary for filling is stored in large rooms. In some asphalt store rooms and filling sheds, the empty and filled drums are transported by rail.

To preserve both asphalt and packaging material, platform conveyors are used in railway transport. This facilitates both loading and unloading. Another means to facilitate loading is to erect a loading ramp elevated in front of the filling units enabling loading the drums into the cars by hand or by some mechanical transport device. Several recommendations and methods have been published for the purpose of cooling and transporting high softening point asphalt qualities not requiring packaging media[36,37,38].

2.9.3.3 *Location of Filling Place*

The units for filling and transport of hot asphalt in bulk do not raise any obstacles in plants producing asphalt. In arranging the filling place, it must be considered whether rail tank cars and tank trucks, or only rail tank cars and only tank trucks will be used for transport. As

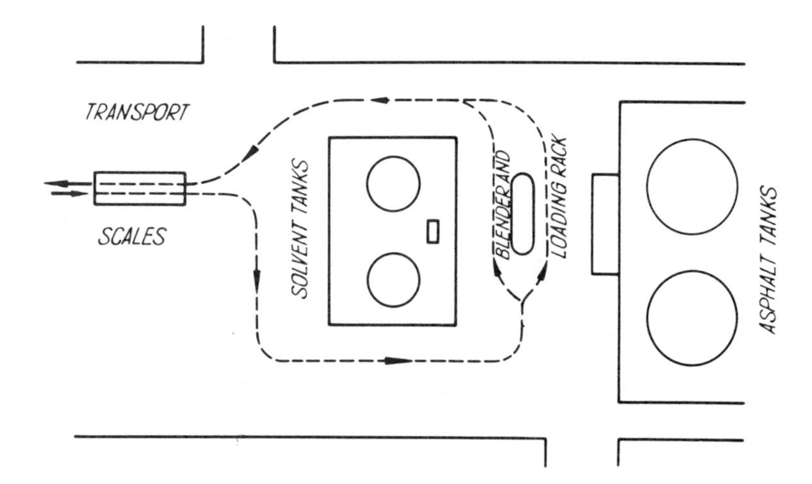

Fig. 46. Tank truck loading place

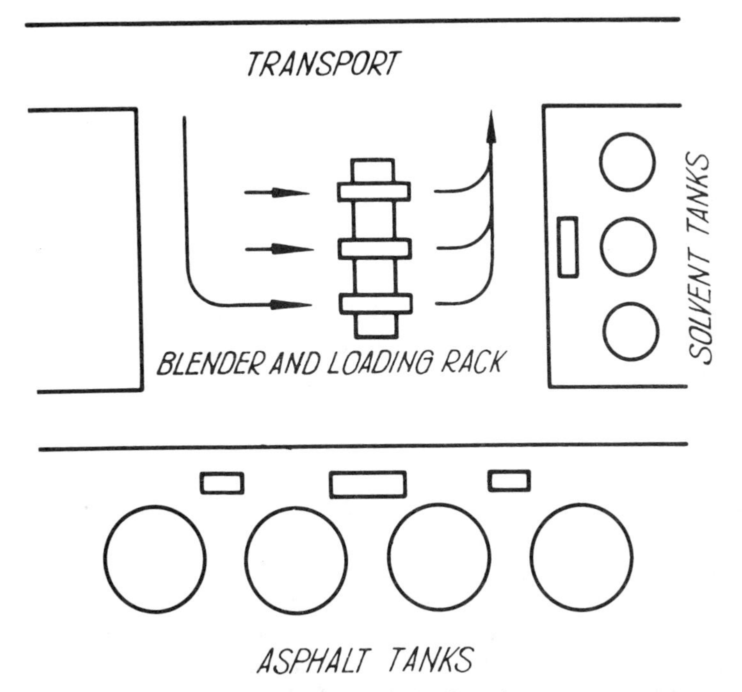

Fig. 47. Tank truck loading place

it has been seen from the data, asphalt in a hot state is chiefly transported by tank trucks. It must be taken into account also when arranging a loading place as to what extent the in line blending method will be utilised in the plant both for asphalt and cut back production and loading.

It is advisable therefore to provide for loading facilities together with the arrangement of the in-line blending system, taking into consideration the suitable selection of the road for the transporting tank trucks.

Some tank truck loading places utilizing in-line blending can be seen in Figures 46 and 47. The necessary asphalt and solvent tanks are also shown in the Figures.

The task is however more difficult if the asphalt producer must also solve the above mentioned problems associated with packaging and filling. The general principles should be mentioned here[39].

1) The filling room ought to be near the blending or mixing and loading tanks, since in the case of elevated tanks pumping can be avoided,

2) A suitable accomodation of the storage during cooling must be

provided for in the filling room. The room should have ample size. In principle, a transit time of at least five days must be stipulated in design.

3) The filling room and the storage place should be covered to enable smooth operation and dry asphalt production. It must also be considered during design work that asphalt consumption changes as a function of the climate prevailing in the individual area. In the hot summer period bulk demands usually rise (road construction) and these must be met. The loading, cooling, and transport capacities to be designed have to be established therefore taking into consideration the satisfaction of these requirements. The size of the room for the filling operation is evaluated on the basis of this.

Various projects have been realised in the individual plants depending on the requirements and on local conditions for the location of the technical parts on the one hand and on transport sections on the other in suitable proportions. Some arrangement possibilities are illustrated in the following Figures.

A loading system joined to a multistage blower is shown in the Figure 48. Upon a certain cooling period the asphalt is loaded in a warm state from the elevated vessel is drained through an opening at the bottom into steel drums or other containers. The drums filled with asphalt are delivered into the storage shed, from which they are delivered into the transport rail tank cars on the other side of the filling room. An alternative for the arrangement of the asphalt, the filling, and the drum units is given in Figure 49. The steel sheets required for drum making arrive for processing directly after unloading them from the rail tank car. The finished drums are stored either in the filling room proper, or

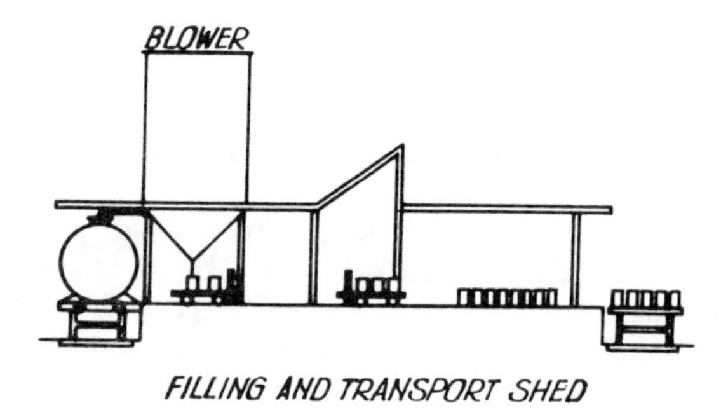

Fig. 48. Loading system connected to a multistage blower

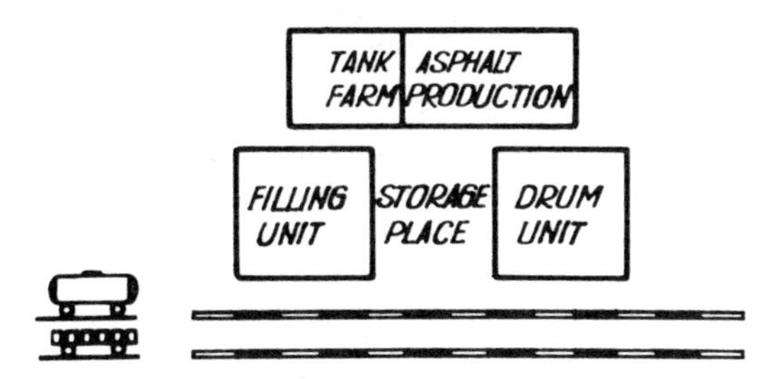

Fig. 49. Arrangement of the asphalt, the filling, and the drum units

in an open space between the room and the drum making unit. The drums are filled at fixed loading points with the asphalt kept in circulation by pumping in the mixing or blowing units, and are then placed in railway tanks upon cooling. The loading system may be used for the loading of tank cars at the same time.

The loading points shown in Figure 50 are situated relatively far from the production units since the plant is equipped for the simultaneous handling of various quality asphalts. An open space is situated near the production units for the filling of straight run or blown asphalts, which are poured into molds or simply drained off. The next filling or storage space is provided for the filling of straight run asphalt or blown asphalt of low softening point into drums. The drum making unit is also erected here. The filling shed farthest from the production unit serves the purpose of finishing distillation asphalt with low

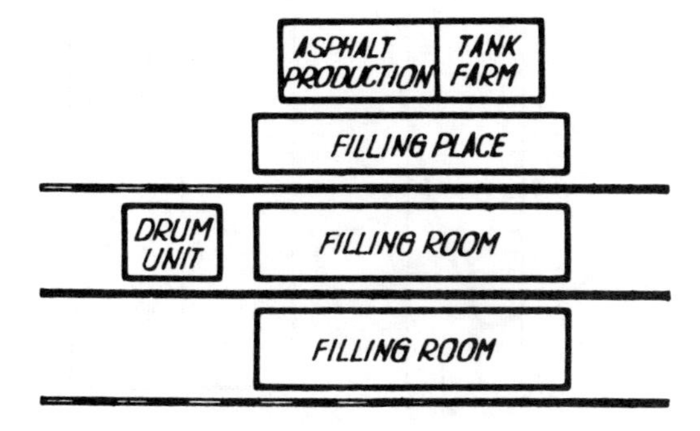

Fig. 50. Asphalt plant equipped for the simultaneous handling of various quality asphalts

softening point. The intermediary storing facilities which are elevated enable loading by pumping as well as filling due to gravity only.

There are several other alternatives beside the above described fundamental solutions. Corresponding to the standpoints mentioned, the purpose of these is the utilization of shorter product lines. The movements of empty drums or those of drums loaded with asphalt, brought-about by hand or mechanically according to the method chosen, must take place on the shortest distance possible in the most economical manner.

As can be gathered from the foregoing, various possibilities may be used for filling and transportation of the asphalt produced in the asphalt plant. Which of these methods is used will depend not only on the asphalt quality but also on local conditions, plant size, and the time of year.

When planning asphalt transportation, the seasonal variations of consumption must always be considered. In the winter months fuel oil is produced in the plants used for asphalt making, while in summer, asphalt is made in capacities corresponding to the asphalt demand. This fact can be seen clearly in Figure 51, illustrating the asphalt delivery of ÖMV (Austria) in the individual months of 1961[40].

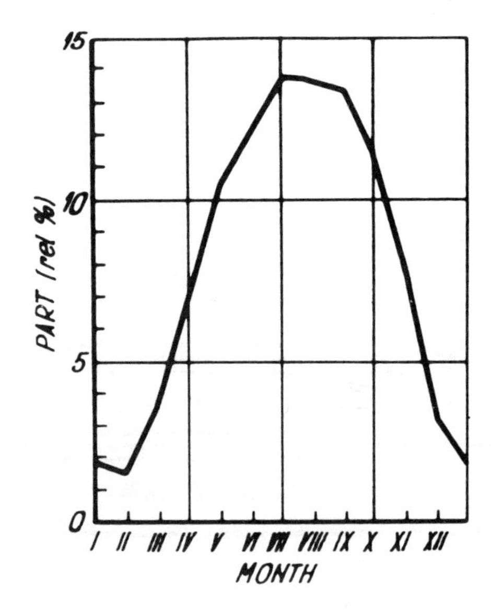

Fig. 51. Seasonal variations in the consumption of asphalt

The most advantageous seems to be the transportation of asphalt in bulk in a hot state, but the consumer must be prepared and equipped suitably for the reception of hot asphalt. If the bulk of asphalt production has to be transported by the plant in a cold solid state either packaged or not, the part of the plant for this purpose will represent the largest part of the asphalt plant[41].

The manpower requirement in an asphalt plant also depends on the method of packaging and that of transportation. Units loading asphalt into rail tank cars or using predominantly cut backs exhibit a minimum manpower requirement, whereas in other plants a staff of more than 1000 people may work. The manpower requirement changes as a function of packaging mechanization.

References

1. Lane, E.C., u. Carton, E. L.: Bureau of Mines, Rep. Inv. 3279 (1935).
2. Prinzler, H.: Einführung in die Technologie des Erdöls. Leipzig 1961. Grundstoffindustrie.
3. Kerényi, E., Zakar, P., MÓZES, Gy.: MÁFKI Közlemények 2. sz. 32 (1961).
4. Gost 912-46.
5. Kerényi, E.: Erdöl u. Kohle, *14*, 611, 703 (1961).
6. Nelson, L., Suresh Patel: Oil Gas J. *62*, febr. 17, 120 (1964).
7. Kerényi, E.: Acta Chim. Acad. Sci, hung. Tomus 31, 113 (1962).
8. Stanfield, K. E., Hubbard. R. L.: Asphalts from Rocky Mountain crude oils, Washington 1949. Bureau of Mines.
9. Le Maire, G. W.: A study of asphalts and asphaltic materials. Colorado 1953. Quarterly. Colorado School of Mines.
10. Zakar, P., Mózes, Gy.: Bitumen, Teere, Asphalte, Peche. *13*, 284 (1962).
11. Nyul, Gy.: Lepárlás. Budapest (1955). Müszaki.
12. Jancsó, T.: Ásványolajipar. Budapest (1956). Müszaki.
13. Lehmann, G.: Erdöldestillation. Mainz (1956). Hüthig und Dreyer.
14. Arutjunow, J., Dh.: Anfahren grosstechnischer Anlagen. Leipzig (1961). Grundstoffindustrie.
15. Roth, H., Mai, H.: Einführung in die Technologie der Schmierölgewinnung aus Erdöl, Leipzig (1962). Grundstoffindustrie.
16. Kraft, W.W.: Ind. Engng. Chem. *40*, 807 (1948).

17. Nyul, Gy., Zakar, R., Mózes, Gy.: Bitumen, Teere, Asphalt, Peche. *10*, 138 (1959).
18. Modern Petroleum Technology. I. P. London (1954).
19. Zakar, P.: Máfki 58. kiadv. Veszprém (1954).
20. Science of Petroleum, London (1938). Oxford University Press.
21. Jackson, J. S.: Petroleum, *8*, 102 (1945).
22. Abraham, H.: Asphalts and allied Substances. New York (1961). D. Van Nostrand.
23. Kastens, M. L.: Ind. Engng. Chem. *40*, 548 (1948).
24. Hoiberg, A. J.: Petroleum Refiner. *26*, 77 (1947).
25. Aixinger, A. O.: Ásványolajtechnológia. Budapest (1951). Nehézipari.
26. Anon: Petroleum Refiner. *28*, 118 (1949).
27. Gagle, D. W.: Oil Gas J. *58*, 121 (1960).
28. Jurina, V.: Bitumen, Teere, Asphalte, Peche. *11*, 50 (1960).
29. Zakar, P.: Magyar Technika. *9*, 440 (1954).
30. Simon, M., Zakar, P.: Máfki 57. kiadv. Veszprém (1953).
31. Néber, E.: A nagylengyeli ásványolaj üzemi lepárlása. Budapest (1954). Mérnōki Továbbképzö Intezet.
32. Zakar, P.: Magyar Kémikusok Lapja, *9*, 150 (1954).
33. Zakar, P., Mózes, Gy.: Acta Chim. Acad. Sci. hung. Tomus *31*, 291 (1962).
34. Nyul, Gy., Zakar, P., Mózes, Gy.: Erdöl u. Kohle *12*, 976 (1959).
35. Freund, M., Zakar, P.: Strassen-u. Tiefbau, *12*, 725 (1960).
36. Zakar, P., Mózes, Gy., Fényi, Gy.-né: Máfki Közlemények 3. sz. 143.
37. Nüssel, H.: Bitumen, Mainz (1958). Hüthig und Dreyer.
38. Hung. Pat. Nr. 152. 637
39. Zakar, P., Mózes, Gy. and others: Máfki 266. kiadv. Veszprén (1962).

2.4.

1. Pfeiffer, J.: The properties of asphaltic bitumen. New York (1950). Elsevier.
2. Nelson, W.L.: Petroleum Refinery Engineering. New York (1949). McGraw-Hill.
3. Brazhnikov, V. T.: Sovremennye ustanovki dlja proizvodstva s. mazocsnyh masel, Moscow (1959).
4. Dunning, H. N., Moore, J. W.: Petroleum Refiner, *36*, 247 (1957).
5. Kádár, I.: Máfki 27. kiadv. Veszprém (1952).
6. Reinkemeyer, L. R.: Oil Gas J. *57*, No. 37. 166 (1959).

7. Kádár, I.: Máfki 139. kiadv. Veszprém (1957).
8. Zakar, P., Mózes, Gy.: Acta Chim. Acad. Sci. hung. Tomus *31*, 281 (1962).
9. Ditman, J. G., Godino, R. L.: Hydrocarbon Processing and Petr. Refiner, *44*, 175 (1965).
10. Oil Gas J. *63*, No. 5. 74 (1965).
11. Themme, T.: Bitumen, Teere, Asphalte, Peche, verwandte Stoffe *16*, 292 (1965).
12. Riedl, W.: Acta Chim. Acad. Sci. hung. Tomus *36*, 461 (1963).
13. Aixinger, I., Száva, N., Vajta, L., Zakar, P.: Oldószeres kenöolajfinomitás. Budapest (1952). Nehézipari.
14. Vámos, E., Zakar, P.: Magyar Kémikusok Lapja. *10*, 114 (1955).
15. Nyul, Gy., Vámos, E., Zakar, P.: Erdöl u. Kohle. *11*, 621 (1958).
16. Kossowicz, L.: Proceedings of the Fith World Petr. Congr. Sec. V. Paper *31*, 411. New York (1959).
17. Zakar, P.: Unpublished communication (1955).
18. Zakar, P.: Unpublished experiment (1965).

2.5.

1. Sachanen, A. N.: The chemical Constituents of Petroleum. New York (1945). Reinhold.
2. Nelson, W. L.: Oil Gas J. March. *2*, 81 (1950).
3. Siemes, W.: Chem.-Ing. Techn. *26*, 479 (1954).
4. Siemes, W., Günther, K.: Chem.-Ing. Techn. *28*, 389 (1956).
5. Calderbank P. H.: Trans. Instr. Chem. Engrs. *36*, 443 (1958).
6. Calderbank P. H.: Moo-Young M. B., Bibby R.: Univ. of Edinburgh.
7. Goppel, J. M., Knoterus, J.: Proc. Fourth World Petr. Congr. Sec. III/G. Rome (1955).
8. Campbell, P. G., Wright, R.: I.E.C. Product Res. Development 3, (1964).
9. Hoiberg, A. J., Garris, W. E.: Ind. Eng Chem. analyt. Edit. *16*, 294 (944).
10. Sergienko, S. R., Delone, I. O., Krasavcsenko, M. I., RUTMAN, L. I. Izv. Turkmen. Adad. Nauk (Ashabad) Serija Fiz. Techn. Him. Geol. Nauk No 3. 10 (1960).
11. Moisickov, S. F., Starobinec, I. S. Himija i Teknologica Topliv i Masel *6* No 9, 41 (1961).
12. Nyul, Gy., Zakar, P., Mózes, Gy.: Erdöl u. Kohle. *12*, 967 (1959).
13. Kossowicz, L.: Proc. Fifth World Petr. Congr. Sec. V. Pap. 31.

14. Pentschev, W.: Acta Chim. Acad. Sci, Hung. Tomus 36. 453 (1963). 16, 292 (1965).
15. Graham, W., Cudmore, W. J., Heyding, R. D.: Can. I. Techn. *30*, 143 (1952).
16. Smith, D. B., Schweyer, H. E.: I.E.C. Process design and development. 2. No. 3. 209 (1963).
17. Greenfeld, S.: I.E.C. Product Res. Development *3*, 158 (1964).
18. Lackwood, D. C.: Petroleum Refiner. *38*, 197 (1959).
19. Senolt, F.: Private Communication (1966).
20. Rescorla, A. R., Forney, W. E., Blakey, A. R., Frino, M. J.: Ind. Eng. Chem. *48*, 378 (1956).
21. Chelton, H. M., Traxler, R. N., Romberg, Romberg, J. W.: Ind. Eng. Chem. *51*, 1353 (1959).
22. Gun., R. B., Gurevitch, I. L.: Novosti neftjanoj tekhniki, No 5. 12 (1958).
23. Murayama, K., Fukushima, T., Fukuda, Y., Shimada, A.: Bulletin of the Jap. Petr. Inst. *2*, 63 (1960).
24. Gundermann, E., Müller, K.: Bitumen, Teere, Asphalte, Peche. *10*, 192 (1961).
25. Gundermann, E.: Chem. Techn. *14*, 324 (1962).
26. Zakar, P., Mózes, Gy.: Erdöl u. Kohle. *14*, 812 (1961).
27. Bitumen, Teere, Asphalte, Peche. *7*, 392 (1956).
28. Klimke, R., Mothes, J., Kohlstrung, G.: Acta Chim. Acad. Sci. hung. Tomus *36*, 499 (1963).
29. Klimke, R., Mothes, J.; Kohlstrung, G.: Chem. Techn. *14*, 725 (1962).
30. Mothes, J., Prinzler, H., Klimke, R.: Chem. Techn. *16*, 716 (1964).
31. Mózes, Gy., Kádár, I., Kristóf, M.: Máfki 324 kiadv. Veszprén (1966).
32. Ropa a Uhlie. *7*, 163 (1965).
33. Rozental, D. A., Filipenko, A. I., Kuznecova, A. S.: Neftepererabotka i Neftehimia (1965) No 2.
34. Böttcher, M.: Bitumen, Teere, Asphalte, Peche, verwandte Stoffe *15*, 134 (1964).
35. Modern Petroleum Technology. J. P. London (1954).
36. Holland, C. J.: Petroleum Engr. *7*, March, 33 (1935).
37. Mihajlov, V. V.: Heftjanije doroz snije Bitumu, Moscow (1949). Doprizdat.
38. Jackson, J. S.: Petroleum. *8*, 82 (1945).

39. Hoiberg, A. J.: Petroleum Refiner. *26*, 77 (1947).
40. Kirk, R. E., Othmer, D. F.: Encyclopedia of Chemical Technology. New York (1948). Interscience.
41. Uhl, W. C.: Petroleum Processing. *5*, 33 (1950).
42. Aixinger, I.: Hungarian Heavy Industries Nr. Twenty-seven, 1. Autumn (1959).
43. Zakar, P.: Bitumen zsebkönyv. Budapest (1961). Mūszaki.
44. Fan-Yun-Nan: Petroleum Processing. *10*, 862 (1955).
45. Ullmann: Encyklopädia der Technischen Chemie, Bd. 4. München (1953). Urban Schwarzenberg.
46. Nelson, W. L.: Oil Gas J. *53*, 158 (1954).
47. Resen, L.: Oil Gas J. *56*, 48, 126 (1958).
48. Pruess, D. B.: Refining Eng. *30*, Oct. C-6 (1958).
49. Purvin, R. L.: Petroleum Engr. *22*. C-29. April (1950).
50. Thurston, Knowles: Ind. Engng. Chem. *28*, 88 (1936).
51. Pfeiffer, J. Ph.: The properties of Asphaltic Bitumen. New York (1950). Elsevier.
52. Nelson, W. L.: Oil Gas J. Oct. *25*, 99 (1947).
53. Khopp, E. P.: Petroleum Processing, *5*, 627 (1950).
54. Nyul, Gy., Mózes, Gy., Zakar, P.: Máfki 127. kiadv. Veszprém (1957).
55. Simon, M., Zakar, P.: Máfki 57. kiadv. Veszprém (1953).
56. Nyul, Gy., Mózes, Gy., Zákar, P.: Máfki 188. kiadv. Veszprém (1960).
57. Kosztrin, K. V.: Neftjanik, *1* No 2. 11 (1956).
58. Gun, R. B., Bakutkin, A. B.: Himija i Tekhnologija Topliv i Masel *3* No 5. 60. (1958).
59. Mózes, Gy., Zakar, P.: Máfki közlemények 2. sz. 193 (1960).
60. Abraham, A.: Asphalts and Allied Substances. New York (1945). D. Van Nostrand.
61. Pauer, O., Harumi, M. M.: Eraöl u. Kohle. *5*, 771 (1952).
62. Shearon, W., H., Hoiberg, A. J.: Ind. Engng. Chem. *45*, 2122 (1953).
63. Sheianu, J., Grigorescu, D., Vasiliu, C.: Petrol si Gaze. *12*, 267 (1961).
64. Mapstone, G. E., Karius, H.: J. Inst. Petroleum. *42*, 361 (1956).
65. Zakar, P., Csikós, R., Mózes, Gy., Kristóf, M.: Magyar Kémikusok Lapja. *18*, 126 (1963).
66. Csikós, R., Mózes, Gy., Zakar, P.: Chem. Techn. *16*, 720 (1964).

67. Csikós, R., Mózes, Gy., Zakar, P.: A fuvatott bitumen. Müszaki Kiadó Budapest (1965).
68. Solc, A.: Nafta *12*, 512 (1962).
69. Gundermann, E.: Erdöl-Kohle-Erdgas-Petrochem. *18*, 780 (1965).
70. Mc Kinney, P. V., Mayberry, M., Westlake, H. E.: Ind. Engng. Chem. *37*, 177 (1945).
71. Siegmann, M. C., by Pfeiffer, J. Ph.: The Properties of Asphaltic Bitumen, 121 p. New York (1950). Elsevier.
72. Van Ufford, 9u. Vlugter, J. C.: Brennstoff-Chemie, *43*, 173 (1962); *46*, 7 (1965).
73. Jurina, V.: Bitumen, Teere, Asphalte, Peche. **11**, 50 (1960).
74. Krenkler, K.: Bitumen, Teere, Asphalte, Peche, verwandte Stoffe *11*, 64 (1960).
75. Krenkler, K.: Bitumen, Teere, Asphalte, Peche. *2*, 105 (1951).
76. Gundermann, E.: Chem. Techn. *11*, 441 (1959).

2.6.

1. Egloff, G., Morell, J. C.: Ind. Engng. Chem. *23*, 679 (1931).
2. Samokovlija, E., Grbac, B.: Nafta. *14*, 283 (1964).
3. Chaix, R.: Rev. Ind. Minerale. *42*, 103 (1960).

2.8.

1. Kastens, M. L.: Ind. Engng. Chem. *40*, 548 (1948).
2. Nyul, Gy., Mózes, Gy., Zakar, P.: Máfki 127. kiadv. Veszprém (1957).
3. Zakar, P.: Máfki 58. kiadv. Veszprém (1954).
4. Bitumen und Asphalt Taschenbuch. Arbit-Bauverlag. Wiesbaden (1957).
5. Senolt, H.: Private Communication. 1965.
6. Zakar, P., Mózes, Gy.: Acta Chim. Acad. Sci, hung. Tomus *31*, 281 (1962).
7. Nyul, Gy., Zakar, P., Mózes, Gy.: Bitumen, Teere, Asphalte, Peche *10*, 138 (1959).
8. Riedl, W.: Acta Chim. Acad. Sci. hung. Tomus *36*, 461 (1963).
9. Pass, F., Schindel, K.: Erdöl-Zeitung. *77*, 456 (1961).
10. Sotir, O., Simionescu, V.: Petrol si Gaze. *11*, 35 (1960).
11. Simpson, W. C., Griffin, R. L., Miles, T. K.: J. Chem. Eng. Data 6, 427 (1961).
12. Benes, V.: Strojirenstvi. *10*, 789 (1960).
13. Barth, E. J.: Refining Engineer. March. C-22 (1958).
14. ASTM Standards on Bituminous Materials. Philadelphia. (1960).
15. Gost 1972-52.

16. Tekhnicseszkie normy na Nefteprodukty, Gostoptehizdat. Moscow (1957).
17. Zakar, P.: Magyar Technika. *9*, 440 (1954).
18. Mihajlov, V. V.: Neftjanüe dorozhnye Bitumu, Moscow (1949). Doprizdat.
19. Zakar, P., Simon, M.: Bitumen, Teere, Asphalte, Peche. *11*, 190 (1960).
20. Zakar, P., Mózes, Gy.: Magyar Kémiai Folyóirat. *64*, 71 (1958).
21. Brooks, K.: Oil Gas J. *59*, March 27., 111 (1961).
22. Traxler, R. N.: Asphalt. New York (1961). Reinhold.
23. Becker, W.: Strasse und Autobahn. *1*, Aug. (1950).
24. Modern road emulsions. Emulsion Road Association. London (1958).
25. Becher, P.: Emulsions, Reinhold. New York, (1957).
26. The Asphalt Handbook, The Asphalt Institute, Maryland (1968).
27. Bituminous Materials in Road Construction, Road Research Laboratory, H. M. S. O. London (1962).
28. Pfeiffer, J. Ph.: The properties of asphaltic Bitumen. New York (1950). Elsevier.
29. Vogt, J. C.: Revue Générale des Routes et des Aérodrames No 385. 93, No 394, 77 (1964) No 400, 101 (1965).
30. Duriez, M.: Revue Générale des Routes et des Aerodromes No 422, 81 (1967), No 428, 90 (1968).
31. Letters, K.: Fette-Seifen-Anstrichmittel. *66*, 112 (1964).
32. Hoiberg, A. J.: Bituminous Materials: Asphalts, Tars and Pithes Vol. II. Interscience, New York (1965).
33. Raudenbusch, H.: Bitumen, Teere, Asphalte, Peche. *13*, 62 (1962).
34. Strassen- und Wegebau mit Bitumenemulsion. Fachverband der Kaltasphaltindustrie. Hamburg (1965).
35. Drukker, I. I., Hoiberg, A. I.: Bituminous Materials Vol II. Interscience. New York (1965), 391.
36. Bitumen emulsification and blending, Hurrell G. C. and Co., Rochester-Prospectus
37. Turbo-Malaxeur "ATOMIX", R. Corlay, Paris. —Prospectus.
38. Super Colloideurs, Moritz R. I. Chatou, France. —Prospectus.
39. Eventov, I. M.: Avtomobil' nye Dorogi *31*. No 1. 14 (1968).
40. Sonic-Application No 295-1, Sonic Eng. Co. Norwalk. —Prospectus.
41. Acid emulsions of bitumen, P ROCHINOR, Auby. —Prospectus.

42. Cationic road emulsions, Armour Hess Chem. Ltd. Westgate. —Prospectus.
43. Les Emulsions de Bitume, Syndicat des Fabr. D'Emulsions Routieres de Bitume, Paris (1966).
44. Seidl, A.: Magyar Epitöipar *16*, 120 (1965).
45. Dinoram, derives cationiques, Auby Prochinor, —Prospectus.
46. Adhesion in road making with Duomeen T, Armour Hess Chem. LTD. —Prospectus.
47. Bewährte und neuentwickelte Zusätze für Bitumenemulsionen Goldschmidt Th. A. G. Essen. —Prospectus.
48. Sauterey, R., Mouton, Y., Giger I., Ramond, G.: Bulletin de Liaison des Laboratoires Routiers No 30, 119 (1968).
49. Ramond, G.: Bulletin de Liaison des Laboratoires Routiers No 30, 157 (1968).
50. Fascicule n° 24, Fourniture de liants hydrocarbones employes a la construction et a l'entretien des chaussees, Cahier des Preseriptions Communes des Marches de Travaux Publies, Paris (1966).
51. Lüder, H., Heerwig, H.: Chem. Techn. *16*, 724 (1964).
52. Agapov, N. F., Zherebcsevskij, V. I., Stavceva, K. V.: Avtomobil' nye Dorogi *27* No 8, 14 (1964).
53. Raikov, K.: Avtomobil'nye Dorogi *28* No 5, 30 (1965).
54. Kejman, V. A.: Avtomobil'nye Dorogi *32* No 1. 19 (1969).

2.9.

1. Shearon, W. H., Hoiberg, A. J.: Ind. Engng. Chem. *45*, 2122 (1953).
2. Sherwood, P. W.: Petroleum, *22*, 427 (1959).
3. Younger, A. H.: Hydrocarbon Processing and Petroleum Refiner. *40*, No 7. 140 (1961).
4. Pruess, D. B.: Refining Engineer. Oct. C-6 (1958).
5. Ney, S., Nagy, S., Helvei, F., Gergály, Gy.: Köolaj és Földgáz *1*. 342 (1968).
6. Spon, H. D.: Erdöl u. Kohle, *10*, 85 (1957).
7. Kastens, M. L.: Ind. Engng. Chem. *40*, 548 (1948).
8. Brooks, K.: Oil Gas J. *59*, March 27., 111 (1961).
9. Anon: Petroleum Processing. *12*, January, 41 (1957).
10. Wishlinski, T. J.: Hydrocarbon Processing and Petroleum Refiner 42, No 7. 157 (1963).
11. Wishlinski, T. J.: Oil Gas J. *61*, Sept. 16. 93 (1963).
12. Brooks, K.: Oil Gas J. *61*, Dec. 2. 134 (1963).

13. In-line mixer, Plenty and Son Ltd. Newbury. —Prospectus.
14. Blending with Plenty, Plenty and Son Ltd. Newbury —Prospectus.
15. Gagle, L. W.: Refining Engineer, March, C-15 (1958).
16. Salusinszky, L., Zakar, P.: Magyar Kémikusok Lapja *11*, 100 (1956).
17. Eckardt, A., W., Becker: Bitumen, Teere, Asphalte, Peche, verwandte Stoffe *16*, 194 (1965).
18. Toers, K.: Erdöl-Z. *75*, 204 (1959).
19. Tanksattelanhänger, Blumhardt, Wuppertal-Vohnwinkel. — Prospectus.
20. Hoiberg, A. J.: Bituminous Materials Vol. II. Interscience, New York (1965). Interscience.
21. Petrol Times. No 67 1715 (1963).
22. Bitumen, Teere, Asphalte, Peche, verwandte Stoffe *14*, 430, 647 (1963).
23. Zakar, P.: Mélyépitéstudományi Szemle *17* 268 (1967).
24. Resen, L.: Oil Gas J. **56**, Dec. 1., 126 (1958).
25. Anon. Petroleum, *29*, 252 (1966).
26. Aixinger, I.: Hungarian Heavy Industries. Nr. twenty-seven 1. Autumn (1959).
27. Rick, A.: Bitumen, Teere, Asphalte, Peche. *3*, 118 (1952).
28. Bitumen und Asphalt Taschenbuch. Arbil-Bauverlag. Wiesbaden, 1957.
29. Goczak, W., Kemplinski, A.: Nafta. *3*, 282 (1954).
30. Thomas, G. O., Wilson, G. A. R.: Revue Belg. des Matieres Plastiques *8* 647 (1967).
31. Modern Petroleum Technology, I. P. London 1954.
32. Hung. Pat. 143.791
33. Zakar, P., Tóth, J.: Freiberger Forschungshefte. A-130, 160 (1959).
34. Dbr Pat. 1118082.
35. Nelson, W. L.: Oil Gas J. May 27., 74 (1943).
36. Csurakov, R. I.: Neftepererabotka i Neftehimia 28 (1963).
37. Rocsev, Iu. R.: Neftepererabotka i Neftehimia 24 (1963).
38. Akimov, V. S., Sajmardanov, N. M.: Neftepererabotka i Neftehimia No 2, 14 (1966).
39. Holland, C. J.: Petroleum Engr. *7*, Febr.-Nov. (1935).
40. Das österreichische Bitumen, Pass, F.: ÖMV Wien.
41. Aixinger, I., Zakar, P.: Freiberger Forschungshefte, A-196, 291 (1961).

3 Uses of Asphalt

With the rapid growth of motor vehicle traffic, the production and application of asphalt for highway use increased at a high rate. Also the other uses are ever increasing together with this. At present the asphalt production all over the world amounts to approximately 50 million ton/year. The average asphalt consumption distribution figures are given for the USA and Western Europe covering the year 1964 in Table 42 (1). It can be gathered from the Table that the demand for asphalt pavement is the largest, civil engineering comes next, and then uses of asphalt in other fields follow. Deviations from the world average values may of course occur fluctuating periodically, corresponding to the conditions prevailing in the individual countries. As for the various production methods, the distribution of USA asphalt production given in Table 43 serves the purpose of orientation[3]. Along with the previous Tables, the distribution of various road asphalts used for asphalt pavement purposes can be seen in Table 44. The 1966 data for the USA and France are shown in the Table illustrating the deviations corresponding to the road construction in each country[3,4]. The French example was included into the Table since there is a relatively large amount of asphalt emulsion production and utilization in the asphalt pavement construction. The road asphalt type proportion naturally exhibits differences for each country.

Parallel to the uses of asphalt, the quality requirements have also

TABLE 42.
Distribution of Asphalt Consumption in the USA and Western Europe in 1964

Application	*USA*	*Western Europe*
Road construction	73.6	79.8
Roofing and building	18.0	20.2
Other uses	8.4	

TABLE 43.
Distribution and Groups of Asphalt Produced in the USA

Groups	*Distribution, %*
Asphalt cements	36
Liquid asphalts and road oils	32
Blown asphalts	24
Other types	8

TABLE 44.
The Distribution of Various Road Asphalts (1966)

	USA %	*France* %
Asphalt cements	66	49
Liquid asphalts	23	19.5
Asphalt emulsions	11	31.5

developed for the individual industries. The standard specifications for paving asphalts are naturally of outstanding importance. Generally, straight run asphalts are used in asphalt pavement construction, although some of the above mentioned production methods are also employed. In the roofing felt industry and in building engineering, considerable quantities of blown asphalts are commonly used. The quality of blown asphalt is, however, not standardized in all the countries, and it is delivered according to marketing conditions and the specifications of the production plants. There are still greater differences between the quality specifications in the various special fields. The different specifications have been developed on the basis of practical experience. However, changes in certain specifications seem to be unavoidable with the advance in asphalt research.

3.1 Reception of Asphalt at the Consumer's

Various units must be erected and certain measures must be taken on the area belonging to the consumer for the purpose of the reception and use of asphalt. The following problems have to be solved when hot asphalt must be received:

a) Unloading
b) Heating
c) Storage

Asphalt transported in rail tank cars may be unloaded by allowing it to flow into deeper containers by gravity, or by pumping from the rail tank car. A maximum 1.5 atm excess air pressure may be used to facilitate filling. Total discharge of asphalt from the rail tank car depends on a definite temperature (viscosity) of the asphalt during unloading. To achieve this condition, the asphalt is loaded at the maker's into the rail tank car at a temperature high enough to enable unloading at the consumer's without any trouble after the time stipulated for the transportation(5). If the asphalt in the rail tank car is not hot enough, it must be heated prior to unloading. Most asphalt tank cars are equipped with steam coils, some with a single-unit coil and some with double-unit coils. The necessary steam lines have to be provided for tank car heating. The specifications covering the tank car heating as well as unloading the asphalt at suitable temperatures (opening and closing of fittings) must be adhered to carefully(6).

In heating all material, it is essential that sufficient steam pressure be used. The time required to heat asphalt so that it can be easily pumped depends entirely on the grade or penetration of the material, the pressure of steam used, and weather conditions; from eight to twelve hours are sufficient under normal conditions.

The minimum temperature to unload the various asphalt types and the minimum vapor pressures necessary for this can be seen in Table 45(7). The values compiled in the Table possess only informative character, since the viscosities and thus the temperatures necessary for reaching identical viscosities fluctuate according to the asphalt

TABLE 45.

Loading Temperatures of Asphalt From Rail Tanks (Temperatures Corresponding to 1500 cSt)

Asphalt	*Designation*	*Temperature (°C)*	*Vapor pressure min. (atm)*
Blown asphalt	85/25	167	10
Blown asphalt	85/40	162	9
Blown asphalt	75/30	155	8
Asphalt cement	B 15	146	6
Asphalt cement	B 25	134	5
Asphalt cement	B 45	123	3.5
Asphalt cement	B 65	117	3
Asphalt cement	B 80	112	2
Asphalt cement	B 200	101	1.5
Asphalt cement	B 300	96	1
Liquid asphalt (DIN 1995)		64	—

origin and method of production. The unloading temperature required may be finally established on the basis of previous viscosity determination of the product. Good railway connections and suitable plant railway sidings may be stipulated generally both in plants and at the consumer's end in the case of transportation in rail tank cars. It is also possible to pump the asphalt direct from one rail tank car into another and to forward it on.

The unloaded asphalt is pumped into suitable storage tanks in plants or at the consumer's. The statements that have been made associated with oil tank farms are valid also here for the construction and operation of these tanks.

The asphalt storage system organization is different in various countries according to the situation of the petroleum refinery. An example of the tank farm at the consumer's is represented here by the local asphalt tanks erected by the French road construction organization. The central tank farm in Rayon Loire et Cher comprises two tanks with a capacity of 1000 m^3 each, four tanks of 45 m^3 each, two with 45 m^3 each, as well as two tanks of 10 m^3 capacities. All of them are heated by steam. Fuel oil heating is used for a similar tank farm on the area at Lot-et-Garonne. Six tank trucks of capacities of 20 m^3 each can be filled in this tank farm simultaneously[8]. The twelve iron concrete tanks having each a capacity of 600 m^3 erected on the area in Maine et Loire department (Angers) are equipped with hot oil system. The temperature measurement and control of the hot oil system and of the asphalt is made by automatic recording devices. The five steel tanks of 750 m^3 capacity in the central storage at Saint Brian, department Côtes du Nord, are loaded with asphalt transported from England by special asphalt tanker. Hot oil is circulated for heating. The central storage area belongs to Shell. Full particulars are given by Nowicki[9] of the French road asphalt storage system including the service system and tank truck transport.

Asphalt transported in tank trucks should be discharged without any difficulty at the consumer's end. The greater mobility of the tank trucks makes it possible to supply the fixed tank farm as well as the various construction sites, which change their location frequently. Also transportable tanks with heating were developed for use at building sites together with the use of tank truck trailers for the transportation of hot asphalt. The volume of the oil-heated transportable binder storage tanks put on the market by various firms amounts to generally from 25,000 to 41000 1b[10]. The asphalt contained in the tanks is warmed by using oil heated in turn by means of oil burners[11].

Beside thermal oil heatings coupled with the tanks, there are also transportable thermal oil generators with heat performances of about 100,000 kcal/hour and above, having a preheating temperature of up to 300°C[12]. Several asphalt storage tanks can be heated by such a generator simultaneously. The thermal oil generator may be utilized also in large mixing tank farms for the heating of lines with oil as a heat transfer medium. All the lines used for the transfer of hot liquid asphalt should have suitable gradients (5%) towards the storage tank or some deep level and have to be equipped with an outlet cock[13].

The heating of asphalts having generally high softening points and transported in the cold state requires melting facilities at the consumers' of these qualities. Prior to treatment the asphalt must be freed from the packaging material, then crushed or ground. There exist various solutions for heating and melting the asphalt transported in a cold state, according to the conditions prevailing in the plants in question. Thus direct heating with coal, coke, gas, and oil may be used or heating with steam or oil as heat transfer media can be used.

Also various electrical heating methods are utilized and recommended for the purpose of heating and melting asphalt.[14,15]. Among the numerous heating methods and units, a unit is also described which serves the purpose of direct heating by flue gas[16].

On removal of packaging material, care must be taken with wrapping paper and plastic foils. Plastic foils melt on heating but in practice clustering often occurs which in turn results in difficulties when filtering the finished product. Humid asphalt or asphalt containing ice must be dried before melting. It is advisable to apply antifoam additives during melting. Heating should be carried out with care with materials having a tendency for foaming.

Vessels with covers are suitable for melting pure asphalt, but if sand or paper contamination occurs, open vessels are more useful. Mixing of the material must be ensured in this case also. A grid on top the melting vessel and a mixing rod will be adequate for this purpose. Coke formation may take place even at low temperatures, therefore the equipment must be cleaned from time to time.

The heated asphalt is commonly pumped into an intermediary tank from the melting unit. The first part of the line should be easily detachable since choking, if any, will take place mostly at this point. The insulation has then to be removed and the material discharged upon heating.

To avoid difficulties arising in the transfer by pumping, Horn[17] recommends the suction of the molten asphalt by vacuum. A unit suitable for this purpose and an appropriate method are also described.

The liquid asphalt obtained from the asphalt transported in a hot state, or by melting the cold asphalt, must be further heated for processing. Depending on local requirements and conditions, the hot oil indirect heating system proved to be best for safety and quality considerations. Electric heating is also used in certain cases. Besides, direct heating similar to the tube still used in the refinery is equally employed here.

3.2 Road Construction

The most important application for asphalt is the road construction. Mixed properly with the aggregate necessary for road building, asphalt is suitable for all kinds of pavements. Besides it is commonly used to prepare the base of modern roads. Asphalt roads are made on the basis of the multilayer system. The prepared subgrade layer is covered by a base course, on which the surface course is laid as a final coating. The early classical base course was a handmade packed layer, which was then gravelled. This was replaced all over the world by a flexible asphalt course, in which mechanically incorporated asphalt gravel or sand asphalt were used. To conform the available aggregates and building materials of various qualities and particle sizes to the different traffic requirements is the task of road construction technologists. The greatest variety of road constructions and pavement types have been developed based on practical experience. These may be distinguished according to various viewpoints. The division used most commonly[18] is that on the basis of the binder:

a) cold (asphalt emulsions or liquid asphalts),
b) warm (liquid asphalt) and
c) hot (asphalt cement),

or according to the method of application

a) spraying and impregnation (surface treatment and penetration macadam),
b) plant mix (mixed macadam, asphalt concrete, mastic),
c) mixed-in-place type.

Open macadam pavements and closed pavements such as asphalt concrete are distinguished according to the particle structure. The various pavement types range from thin coatings to the thickest coating layers according to the volume of traffic.

The individual problems arising in road construction will not be dealt with here; it should, however, be mentioned that the preparation and structure of an asphalt pavement inevitably requires compromises[19]. The economic result of meeting the technical requirements under the prevailing circumstances depends on the right selection of the pavement. A fundamental standpoint is, of course, the choice of the most suitable asphalt for the construction of a satisfactory and durable pavement, but this is not the only condition. All the circumstances must be considered under which the road will be used during its application to develop the right technique and to select the most suitable asphalt quality. To achieve the right viscosity, the asphalt is heated during treatment in large quantities and then contacted with the hot mineral aggregate in a thin layer. The asphalt in the asphalt layer formed on the pavement depends on the local weather conditions, high summer and low winter temperatures, as well as any other weather effects, such as the fluctuations in the daily temperatures.

In addition there is the mechanical stress due to traffic. The above mentioned stress must be emphasized in judging paving asphalt qualities, since no substantial changes ought to take place in the quality of a suitable paving asphalt due to these factors either during treatment or during construction. The quality specifications for paving asphalt have been standardized in most countries. The specifications exhibit certain deviations in the individual countries.

Asphalt binders for paving are:

a) Paving asphalt or asphalt cement
b) Liquid (cut back) asphalt
c) Asphalt emulsions

3.2.1 Paving Asphalt

Paving asphalts are classified according to their penetrations measured at 25°C. The classifications commonly used in the individual countries differ from each other.

The most important specifications of U.S., French, German, and Soviet standards on paving asphalts in the individual countries are shown in Tables 46, 47, 48, and 49. On comparison of the data compiled in these Tables, the different standardizations can be observed. The common purpose of these is to determine the most important characteristics of the groups under the different conditions prevailing

TABLE 46.
Specifications for Asphalt Cements (The Asphalt Institute)

Characteristics	*AASHO Test Method*	*ASTM Test Method*	*Grades*				
Penetration, at 25°C 100 g., 5 sec.	T 49	D 5	40–50	60–70	85–100	120–150	200–300
Viscosity at 135°C Kinematic, Centistokes	T 201	D 2170	240+	200+	170+	140+	100+
Saybolt Furol, SSF	......	E 102	120+	100+	85+	70+	50+
Flash Point (Cleveland Open Cup), °C	T 48	D 92	232+	232+	232+	218+	177+
Thin Film Oven Test	T 179	D 1754	......	......	......	......	......
Penetration After Test, 25°C 100 g, 5 sec, % of Original	T 49	D 5	55+	52+	47+	42+	37+
Ductility							
At 25°C, cms.	T 51	D 113	100+	100+	100+	60+	......
At 15.6°C, cms.			......	......	......	......	60+
Solubility in Carbon Tetrachloride, %	T 44	D 2042	99.0+	99.0+	99.0+	99.0+	99.0+

General Requirement—The asphalt shall be prepared by the refining of petroleum. It shall be uniform in character and shall not foam when heated to 177°C.

TABLE 47.
French Asphalt Specifications C.P.C. Fascicule n° 24 (1967)

Designation	*180/200*	*100/120*	*80/100*	*60/70*	*40/50*	*20/30*
Softening point, °C	34–43	39–48	41–51	43–56	47–60	52–65
Penetration at 25°C, 0.1 mm	180–200	100–120	80–100	60–70	40–50	20–30
Density at 25°C	1.00–1.07	1.00–1.07	1.00–1.07	1.00–1.10	1.00–1.10	1.00–1.10
Loss on heating, (5 hr at 163°C), % max.	2	2	2	1	1	1
Retained penetration after loss on heating, % min.	70	70	70	70	70	70
Flash point, (Cleveland), °C min.	230	230	230	230	250	250
Ductility at 25°C, cm, min.	100	100	100	80	60	25
Solubility in CS_2, % min.			99.5			
Paraffin wax content, % max.			4.5			

TABLE 48.
Road Asphalt (DIN1995)

Designation	*B-300*	*B-200*	*B-80*	*B-65*	*B-45*	*B-25*	*B-15*
Penetration at 25 °C, 0.1 mm	250–320	160–210	70–100	50–70	35–50	20–30	10–20
Softening point, °C	27–37	37–44	44–49	49–54	54–59	59–67	67–72
Fraass breaking point, °C	−20	−15	−10	−8	−6	−2	+3
Cyclohexan insoluble, max. %			0.5				
Ductility							
at 15 °C, cm	100	—	—	—	—	—	—
at 25 °C, cm	—	100	100	100	40	15	5
Paraffin wax, max. %			2.0				
Loss on heating (163°C, 5 hr) max. %	2.5	2.0	1.5	1.0	1.0	1.0	1.0
Penetration decrease after loss on heating test, max. %	60	60	60	60	60	50	40

TABLE 49.
Improved Road Asphalt (GOST 11954-66)

Designation	*BND 200/300*	*BND 130/200*	*BND 90/130*	*BND 60/90*	*BND 40/60*
Penetration at 25 °C, 0.1 mm	201–300	131–200	91–130	61–90	40–60
at 0°C, 200g 60s, min.	45	35	28	20	13
Ductility at 25°C cm, min.	—	65	60	50	40
Softening point, °C min.	35	40	45	48	52
Fraass breaking point max. °C	−20	−18	−17	−15	−10
Adhesion on marble and sand		suitable			
After loss on heating (5 hr 160°C)					
Penetration of residue, % of original, min.	—	60	70	80	80

in each country. In connection with U.S. specifications, it should be mentioned that the development of design and construction standards, including material specifications for State highways, is primarily the responsibility of the individual State highway departments.

A coordinated effort to develop a guide for standard material specifications for use by all States is the responsibility of the American Association of State Highway officials (AASHO). Specifications for highway construction materials also are promulgated by the American Society for Testing Materials (ASTM) and by the General Services Administration (GSA) of the Federal Government. Another institution occupied with the development of paving asphalt specifications is the Asphalt Institute. The Asphalt Institute is an international, non-profit association sponsored by members of the petroleum asphalt industry to serve both users and producers of asphaltic materials through programs of engineering, service, research and education.

The specifications issued by the Asphalt Institute are not obligatory national specifications, but are in widespread use. Table 46 contains the specification for asphalt cements as compiled by The Asphalt Institue. The specifications of the Table include the most important changes recommended in this field during the last decade. These are the decrease in the number of penetration groups, the specification of the kinematic viscosity, and the introduction of the Thin Film Oven Test method.

Comparative results of the new American method for the Thin Film Oven Test are given by Nitsch[20] surveying the new specifications and recommendations in the U.S.A. The Thin Film Oven Test has replaced the Loss on Heating Test. The Thin Film Oven Test is used to obtain a general indication of the amount of hardening which may be expected to occur in an asphalt cement during the plant mixing operation. This hardening tendency is measured by penetration tests made before and after the Thin Film Oven Test. The Asphalt Institute considers that the change in weight of the asphalt cement during the Thin Film Oven Test is of no real significance and therefore such requirements are not included in its specifications. The Thin Film Oven Test introduced in the USA has the same results as those of the corresponding DIN specification.

The following statements can be made on the basis of comparing the road asphalt specifications shown in the Tables. The greatest number of groups can be found in the DIN standard specifications including lower penetration qualities. Only the GOST specifications form a continuous transition among them. The softening point limits are given together with penetration groups in the German and French standard specifications. Up to now the viscosity value is specified only in the U.S. standard. The Soviet standard covers adhesion tests also. Allowable paraffin wax content is specified in the French and German standards. The determination of Fraass breaking point is included in the German and the Soviet standards, and only the Soviet standard shows the penetration value required at 0°C in each group. The ductility requirements are lower in the Soviet standard than those of the other standards. A lower Fraass breaking point value is stated in the DIN standard as for asphalts obtained from German crude oils. The least penetration decrease is allowed by the French standard after the loss on heating test. Small differences in the individual tests must also be considered during comparison. However, the discussed deviations throw light on the situation prevailing in the individual countries as regards asphalt binder specifications. It is obvious that there is a general trend all over the world to develop better standard specifications.

Research work is chiefly carried out from two standpoints related to one another. On the one hand correlating the existing specifications with road construction experience and more accurate definitions are aimed at. On the other hand the differences arising from the various crude oil qualities utilized in asphalt production and from the various production techniques used should be clarified to ensure the manufacture of good quality road asphalts. It is necessary for reasonable results that part of the possible variables must be eliminated and research work has to be carried out on the basis of previous results. Due to the great number of factors and their different variations, the task is very intricate. It is clear that in spite of increasing research work most of the problems have not yet been solved and require further investigations.

The research problems covering the above are being taken up in various conferences(21,22) and the problems and recommendations are submitted by the researchers according to local problems or their research activities. Stinsky(23,24) carried out detailed investigations on road asphalts available in Switzerland. The correlations between asphalt production and quality specifications are taken up on the basis of samples belonging in the same group but obtained from various firms and at different times. A revision of the specifications is recommended by him to ensure a more uniform quality road asphalt for the users. It is emphasized by Zakar(25) that the possibility of making available uniform road asphalt quality by specifying a certain crude oil and a definite production method can only be used in a comparatively small area. The Hungarian road asphalt standard specifies the Nagylengyel crude oil of high asphalt content as a raw material and the vacuum distillation method.(26). The problems arising are much more intricate when different type crude oils are processed since distillation residues, propane asphalt, lubricating oil extracts, and various oil distillates may be used as blending components, as well as various feed stocks may be blown. Difficulties encountered in the standard specifications for road asphalt from Romashkino crude oil are referred to by Zakar(27). Permanent uniform quality product cannot be guaranteed safely in every respect by the conventional standard specifications. Mihailov(28) investigated the influence of production techniques on asphalt quality. The performed investigations have shown that the use of different production techniques with asphalts from the same crude oil leads to final products differing considerably in chemical composition and consequently in physical and mechanical properties. Kalbanovskaja(29), analyzing the asphaltene, resin, and oil contents of asphalts obtained from different crude oils by different technologies, has suggested dividings asphalts into three groups:

Asphalts of Type 1 Contain more than 25% asphaltenes, less than 24% resins, and above 50% oil hydrocarbons,

Asphalts of Type 2 contain less than 18% asphaltenes, more than 36% resins and less than 48% oil hydrocarbons,

Asphalts of Type 3 contain 20-23% asphaltenes, 29-34% resins, and 46-50% oil hydrocarbons.

Comparison of all these asphalts as construction materials has shown that preference should be given to asphalts of Type 3. These asphalts possess all the advantages of asphalts of Type 1 and 2, whereas the disadvantages are far less.

The asphalt with a predetermined chemical composition and properties corresponding to the optimal structure can be produced by a given technological process selecting suitable crude oils, and from various crude oils by choosing different techniques. The above studies form the basis of the standard GOST 11954-66. Since 1966 the refineries of the Soviet Union have begun the huge production of asphalts of type 3, corresponding to the requirements of this standard.

It is obvious that, in asphalt pavement construction, the question arises from time to time whether the early quality specifications still guarantee an adequate security from the standpoint of service requirements. At present a general trend of checking all the previously developed specifications all over the world is manifest, involving the stating of the characteristic data of good quality paving asphalts. It is well known that asphalts in agreement with the standard specifications exhibit different properties during subsequent tests. Endeavours of paving technologists are therefore fully justified to determine the qualities of asphalts better meeting the requirements.[30]. First of all asphalt viscosity measurement and specification are recommended by different researches[31,32] besides qauality characteristics used up to now. Thus Traxler suggests[33] the determination of the absolute viscosity instead of the determination of the softening point and the penetration commonly used at present. He stresses that the utilization of asphalt binder under service conditions involves a suitable viscosity, and thus the determination of the upper and lower viscosity limits is highly desirable. The aging phenomena should be considered specially. Ewers[34] calls the attention of researchers to the stress under service conditions and emphasizes that the behavior of all road binders depends not only on the properties of the binder during construction, but also on the changes of binders taking place in use as a function of time.

It was stated, based on test results obtained in the USA in the evaluation of several experimental roads that there exists a correlation between the ductility of an asphalt that has undergone rapid aging in a heating test, and the useful life of the pavement. Associated with the above statement, Krenkler[35] reports on ductility tests carried out at low temperature and the corresponding testing specifications.

Considering all the requirements Krom and Dormon recommend in the suggested specification[36] for road asphalt firstly the determination of viscosity at 60°C or EVT and Stiffness at 0°C and at −10°C. The Fraass Breaking Point (max.) in conjunction with a minimum penetration at 15°C may be used as a provisional measure until suitable routine tests for direct stiffness measurements are available. Until then at EVT (10^4 s) would also be needed for characterization. An important aspect of this form of specification is its flexibility. The viscosity at maximum road temperature (60°C) and the low temperature stiffness, or their alternatives may be varied to suit local requirements. By repeated recommendations, Krom and Dormon endeavoured setting up internationally accepted criteria for the performance require-

TABLE 50.
Critical Conditions for Asphalt in Road Mixes

	Critical condition		*Significant property of asphalt in the mix*
Behaviour in	*Temperature*	*Time of loading (s)*	
Application:			
Mixing	High (well above 100°C)	—	$EVT_{200\text{-cs}}$ min and max
Spreading	High	—	
Compaction	High	—	$EVT_{20,000\text{-cs}}$ min and max
Service:			
Plastic deformation	High road temperature	Long	Viscosity* min
Fatting up	High road temperature	Long	Viscosity* min
Cracking:			
Traffic stresses	Low road temperature	Short (10^{-2})	Stiffness modulus** max
Thermal stresses	Low road temperature	Long	Stiffness modulus** max
Fretting	Low road temperature	Short (10^{-3})	Stiffness modulus** max

Note: As discussed the asphalt should in principle be in a condition representative of changes which may be occurred: * during the mixing process; and ** after ageing in service.

TABLE 51.
Alternative Form of Specification

Properties	*Requirements*	*Notes*
$EVT_{200\ cs}$ (mixing temperature)	Min and max	1
Properties after ageing test at mixing temperature:		2
Viscosity at 60°C	Min and max	
Properties after further ageing:		3
Stiffness modulus 0°C, 50 c/s	Max	4
Stiffness modulus 10°C, 10^4 s	Max	5
General requirements:		6
Flash point (Coc)	Min	
Specific gravity		
Ash	Max	

Notes:
1. Grade designation, e.g. 150+5°C.
2. For the time being we consider the thin film oven test (ASTM D 1754), loss on heating (DIN) and rolling thin oven test (Hveem) equally suitable for the purpose.
3. The degree of further ageing, e.g. in the microdurability test, depends on the road construction. In the case of dense carpets no further hardening may be necessary and tests can be carried out on the sample aged at mixing temperature.
4. Stiffness modulus 0°C, 50 c/s can be substituted by Fraass breaking point.
5. Stiffness modulus −10°C, 10^4 s can be approximated by penetration at 15°C or 25°C.
6. These items have no mandatory limits.

ments of road asphalts on the basis of their mechanical properties and the changes in these properties during application and service. Such changes will arise as a result of traffic and under the influence of air, light and water. Asphalt critical conditions in road mixes are compiled in Table 50. There is no doubt that rheological tests on aged binders simulating the ageing which occurs during mixing and during service are necessary. The properties of the binder in the paving mix rather than the binder as supplied are the important properties. Considering the questions involved, the suggested specifications are summarized in Table 51 in alternative form as a basis for discussions[37]. Current specification tests to measure flow properties are empirical and requirements have been developed on a trial and error basis. Ivanov[38] studying quality requirements of asphalts in relation to local conditions (climate, traffic) states that specifications for asphalt in general include several properties having no direct link with its behaviour in the road. Research and development studies are under way in the United States to develop fundamental tests and knowledge to define the essential engineering properties that can be used in specifications. The specifications suggested by Krom and Dormon and the "Research Specification" proposed in the United States appear to offer a sound basis for continuing research directed towards characterizing asphalt cements by the use of fundamental knowledge.

Welborn[39], considering the above circumstances, states that the literature contains many references on the properties of asphalt, but information is lacking on the relations of these properties to pavement performance.

During the past decade, the engineering, design, and construction of asphalt pavements have advanced from the art which depended largely upon the experience of local engineers to an engineerings cience. Today there is a considerable research activity in this field with strong indications that success is imminent. Izatt[40], stating the above conditions, emphasizes that during the past decade new engineering procedures for the construction of asphalt pavements and new methods for quality control have been developed. Progress in this direction is an absolute necessity. Monsmith[41] dealing with the function of asphalt layers in pavements with regard to performance, summarizes design considerations for asphalt pavements and other asphalt uses also. The summary emphasizes the complex nature of pavement and asphalt mix design, and indicates that such design requires consideration of a series of factors. In addition it emphasizes the point that each asphalt mixture must be tailored to the specific conditions of environment and loading and that each design must attempt in view of the mixture and the pavement structure to strike a balance among a series of complex and at times interrelated factors.

Associated with the standard specifications, the necessity of strict consideration of highway construction technical standpoints is stressed by Zakar and Simon[42]. The behavior and the evaluation of paving asphalt under actual service conditions is treated by Güsfeldt[43]. The general trend of removing conventional methods and developing determinations of values in the cgs-system is unavoidable in the long run.

The main purpose of these discussions and experiments is the development of good pavements. Beside asphalt, however, a suitable aggregate and correct road construction techniques are required. Increased control of all the processes is also necessary for unobjectionable road constructions. Upon laboratory tests carried out on various asphalt types, all the practical experiments concerning asphalt quality will give reliable data only if the road construction techniques have been applied correctly and all the phenomena observed subsequently are only due to the asphalt binder. Of course, these higher

TABLE 52.
Suggested Grades of Asphalt Cement for Different Climates

Paving Uses	Climate	
	Hot and Temperature	Cold
Airfields		
Runways	60–70	85–100
Taxiways	60–70	85–100
Parking Aprons	60–70	85–100
Highways		
Heavy Traffic	60–70	85–100
Medium to Light Traffic	85–100	120–150
Streets		
Heavy Traffic	60–70	85–100
Medium to Light Traffic	85–100	85–100
Driveways		
Industrial	60–70	85–100
Service Station	60–70	85–100*
Residential	85–100	85–100
Parking Lots		
Industrial	60–70	60–70
Commercial	60–70	85–100
Recreational		
Tennis Courts	85–100	85–100
Playgrounds	85–100	85–100
Curbing	60–70	85–100

* 70-70 penetration normally used for sheet asphalt.

quality requirements cannot be stricter than defects that may occur in road construction and are still within the limits of tolerance[44].

The Asphalt Institute has produced a chart for indicating the desired penetration grade asphalt to be used for a particular application, climatic conditions, and traffic requirements. The suggested grades of asphalt cement are shown in Table 52.

3.2.2 Liquid Asphalts

Volatile petroleum solvents will cut back asphalts to a relatively fluid state by lowering the viscosity. The workability is thereby improved when cold aggregates and limited equipment are used for certain types of road construction. Upon curing the evaporation of the solvent the remaining asphalt cement will be in approximately the same condition as before being taken into solution.

Low viscosity cut back asphalts can be used favorably in certain road construction techniques. Various changes in this have taken place in the specifications of several countries during recent years. The term liquid asphalt has been generally accepted, and is also included in standards. As a basis for classification the solvent evaporation time is used.

On December 6, 1961, the Asphalt Institute adopted new specifications for all RC (Rapid-Curing), MC (Medium-Curing), and SC (Slow-curing) grades of liquid asphalts. These new grades are classified on the basis of kinematic viscosity at 60°C. They replace the grades formerly classified on the basis of Saybolt Furol viscosity at several temperatures. The number in the grade designation system signifies the lower limit for the kinematic viscosity at 60°C. Thus an MC-70 has a minimum kinematic viscosity at 60°C of 70 centistokes. The upper limit of the viscosity range is twice the lower limit. These new grades have since been adopted by nearly all States and roadbuilding agencies in the United States and by many agencies in Canada.

The Asphalt Institute specifications for liquid asphalts[45] are compiled in Tables 53, 54, and 55.

The types "common" and rapid curing are distinguished for liquid asphalts in the new French specifications. Separate types are fixed for products diluted with tar oils. The Standard Tar Viscosimeter with 4 and 10 mm orifice is used to determine the viscosity of French liquid asphalt types. The test is carried out at 25°C. The usual groups are:

Very fluid	0-1
Fluid	10-15
Middle Viscosity	50-100
Viscous	150-200
Very viscous	400-500

Only two groups are shown in the rapid-curing types, 25-75 and 100-250.

The Standard specification covering liquid road asphalts issued in the Soviet Union previously (GOST 1972-52) relates to medium and slow curing types. The viscosity measured at 25°C and 60°C with STV of 5 mm orifice serves as a basis to distinguish six groups. The improved road liquid asphalts technical requirements (GOST 11955-66) also include only medium and slow curing types. The basis of this specification is uniformly the viscosity measurement at 60°C with STV having an orifice of 5 mm. The medium curing type is produced from the BND 60/90 penetration asphalt group according to GOST 11954-66 by dilution. The slow curing type is made from crude oil residues and suitable diluents. The standard contains a specification for adhesion

TABLE 53.
Specifications for Rapid-Curing (RC) Liquid Asphalts (The Asphalt Institute)

Characteristics	AASHO Test Method	ASTM Test Method	Grades RC–70	RC–250	RC–800	RC–3000
Kinematic Viscosity at 60°C cs.	T 201	D 2170	70–140	250–500	800–1600	3000–6000
Flash Point (Tag. Open Cup), °C	T 79	D 1310	......	26.7+	26.7+	26.7+
Distillation						
Distillate (% by Volume of Total Distillate to 360°C)						
To 190°C			10+	......	......	......
To 225°C	T 78	D 402	50+	35+	15+	......
To 250°C			70+	60+	45+	25+
To 316°C			85+	80+	75+	70+
Residue from Distillation to 360°C, % Volume by Difference			55+	65+	75+	80+
Tests on Residue from Distillation						
Penetration, 25°C, 100g, 5 sec.	T 49	D 5	80–120	80–120	80–120	80–120
Ductility, 25°C	T 51	D 113	100+	100+	100+	100+
Solubility in Carbon Tetrachloride, %	T 44	D 2042	99.5+	99.5+	99.5+	99.5+
Water, %	T 55	D 95	0.2−	0.2−	0.2−	0.2−

General Requirement—The material shall not foam when heated to application temperature recommended by The Asphalt Institute.

TABLE 54.
Specifications for Medium-Curing (MC) Liquid Asphalts (The Asphalt Institute)

Characteristics	*AASHO Test Method*	*ASTM Test Method*	*Grades*				
			MC–30	*MC–70*	*MC–250*	*MC–800*	*MC–3000*
Kinematic Viscosity at 60°C	T 201	D 2170	30–60	70–140	250–500	800–1600	3000–6000
Flash Point (Tag. Open Cup), °C., cs.	T 79	D 1310	37.8+	37.8+	65.6+	65.6+	65.6+
Distillation							
Distillate, % by Volume o Total Distillate to 360°C							
To 225°C			25–	20–	0–10	......	......
To 260°C	T 78	D 402	40–70	20–60	15–55	35–	15–
To 316°C			75–93	65–90	60–87	45–80	15–75
Residue from Distillation to 360°C, % Volume by Difference			50+	55+	67+	75+	80+
Test on Residue from Distillation							
Penetration, 25°C, 100g, 5 sec.	T 49	D 5	120–250	120–250	120–250	120–250	120–250
Ductility, 25°C, cms.	T 51	D 113	100+	100+	100+	100+	100+
Solubility in Carbon Tetrachloride, %	T 44	D 2042	99.5+	99.5+	99.5+	99.5+	99.5+
Water, %	T 55	D 95	0.2−	0.2−	0.2−	0.2	0.2−

General Requirement—The material shall not foam when heated to application temperature recommended by The Asphalt Institute.

TABLE 55.
Specifications for Slow-Curing (SC) Liquid Asphalts (The Asphalt Institute)

Characteristics	*AASHO Test Method*	*ASTM Test Method*	*Grades*			
			SC–70	*SC–250*	*SC–800*	*SC–3000*
Kinematic Viscosity at 60°C, cs.	T 201	D 2170	70–140	250–500	800–1600	3000–6000
Flash Point (Cleveland Open Cup), °C	T 48	D 92	65.6+	79.4+	93.3+	107.2+
Distillation						
Total Distillate to 360°C, % by Volume	T 78	D 402	10–30	4–20	2–12	5–
Kinematic Viscoity at 60°C Stokes	T 201	D 2170	4–70	8–100	20–160	40–350
Asphalt Residue of 100 Penetration, %	T 56	D 343	50+	60+	70+	80+
Dustility of 100 Penetration Residue at 25°C, cms.	T 51	D 113	100+	100+	100+	100+
Solubility in Carbon Tetrachloride, %	T 44	D 2042	99.5+	99.5+	99.5+	99.5+
Water, %	T 55	D 95	0.5–	0.5–	0.5–	0.5–

General Requirement—The material shall not foam when heated to application temperature recommended by The Asphalt Institute.

test also apart from the commonly used qualifying and classifying tests. The maximum quantity of water soluble compounds amounts to 0.3%, providing for asphalts containing surface active agents a maximum of 0.6%.

A new statement is made in the above specification; that liquid asphalts may be made either with surface active agents or without them. If liquid asphalts meet the adhesion requirements by themselves, they can made without additives. A suffix "p" is applied to mark qualities containing additives. Thus one of the groups containing additives according to GOST 11955-66 Soviet Standard is marked SzGp-15/25.

The Hungarian liquid asphalts are made in the production plant with additive according to standard. The standard specifications also contain requirements for the adhesion value[47]. It seems to be necessary to state here that one of the principal functions of an asphaltic binder is to act as an adhesive either between road aggregates or between aggregate and the underlying road surface. There are occasions when the adhesive bond may fail and the durability of the asphaltic material may be seriously affected.

Failure of a bond already formed is commonly referred to as "stripping" which is brought about by the displacement of the asphaltic binder from the aggregate surface by water[48]. The meaures taken to improve adhesion have to be mentioned associated with liquid asphalts. In the mix asphalts the heated and dried mineral aggregates are always coated with the binder in layers with uniform thickness by means of mechanical operations. Adhesion arises here as a relatively slight problem. The asphalt binder is however sprayed on the cold aggregate surface in the pavements made by spraying. The binder is spread on the aggregate only after some days or even some weeks due to the action of hot weather and the sun on the one hand, and on the other the kneading effect of traffic. The aggregate being wet, it is inevitable that the air humid coarse layer becomes wet during such a long time. The use of adhesion improving substances ensures constant adhesion of the asphaltic material on the aggregate in a humid atmosphere or even in rain. Any number of investigations have been carried out to select suitable additives for the purpose of improving adhesion. As a result many compounds proved to be satisfactory[49,50,51]. Numerous adhesion improvers are recommended by the producers in the individual countries. A more promising approach is to employ additives which show cationic surface activity[52,53,54].

The following compounds were satisfactory:

1) Long chain aliphatic amines or polyamines containing at least 15 carbon atoms
2) Cyclic nitrogen compounds with long aliphatic chains and benzene rings
3) Quaternary ammonium compounds with at least one aliphatic chain on the nitrogen atom

The Soviet standard GOST 11955-66 dealt with above on improved road liquid asphalt contains recommendations on additives for use. These are essentially materials belonging to the above groups. They tend to be strongly adsorbed at the aggregate/binder interface and thereby reinforce the adhesive bond and render it resistant to water.

The quantity of the necessary agent depends on the aggregate, the agent type, the atmospheric conditions, and traffic circumstances. Higher temperatures are disadvantageous for addition of the additive, thus it is added generally prior to use. It should be mentioned that good adhesion can also be achieved by a suitable selection of tar oils or their derivates.

3.2.3 Asphalt Emulsions

Large quantities of asphalt emulsions are also used for road pavement construction in certain countries. The number of standardized asphalt emulsion grades depends on the local consumption and the prevailing circumstances and is different in the individual countries.

Asphalt emulsions are classified according to various viewpoints. They may be distinguished by the charge of the asphalt particles, anionic or cationic.

It has been mentioned in the foregoing that the production and use of cationic emulsions is ever increasing all over the world. Apart from this, anionic emulsions are also used in the individual countries in different proportions. Formerly quality specifications were developed for anionic emulsions corresponding to production conditions. Quality specifications on cationic emulsions have recently appeared, however, in many countries.

Asphalt emulsions with 50, 55, 60, 65, and 70% asphalt contents are distinguished by the French standard specification[55]. Asphalt emulsions may be classified on the basis of their viscosities measured with the Engler viscosimeter as follows:

Fluid 2-6	with a viscosity of 2-6 E
Semi fluid 6-15	with a viscosity of 6-15 E, and
Viscous 15-30	with a viscosity of 15-30 E.

TABLE 56.
Specifications for Anionic Emulsified Asphalts (The Asphalt Institute)

			Grades				
			Rapid Setting		*Medium Setting*	*Slow Setting*	
Characteristics	*AASHO Test Method*	*ASTM Test Method*	*RS–1*	*RS–2*	*MS–2*	*SS–1*	*SS–1h*
Tests on Emulsion							
Furol Viscosity at 25°C, sec			20–100	......	100+	20–100	20–100
Furol Viscosity at 50°C, sec			......	75–400	......	......	......
Residue from Distillation, % by Weight			57+	62+	62+	57+	57+
Settlement, 5 days, % Difference	T 59	D 244	3−	3−	3−	3−	3−
Demulsibility							
35 ml. of 0.02 N $CaCl_2$, %			60+	50+	......	......	......
50 ml. of 0.10 N $CaCl_2$, %			......	......	30−	......	......
Sieve Test (Retained on No. 20),			0.10−	0.10−	0.10−	0.10−	0.10−
Cement Mixing Test, %			......	......	......	2.0−	2.0−
Test on Residue							
Penetration, 25°C, 100 g., 5 sec	T 49	D 5	100–200	100–200	100–200	100–200	40–90
Solubility in Carbon Tetrachloride, %	T 44	D 2042	97.5+	97.5+	97.5+	97.5+	97.5+
Ductility, 25°C, cms.	T 51	D 113	40+	40+	40+	40+	40+

TABLE 57.

Specifications for Cationic Emulsified Asphalts (The Asphalt Institute)

Characteristics	AASHO Test Method	ASTM Test Method	Grades					
			Rapid Setting		*Medium Setting*		*Slow Setting*	
			RS–2K	*RS–3K*	*SM–K*	*CM–K*	*SS–K*	*SS–Kh*
Test on Emulsion								
Furol Viscosity at 25°C, sec	T 59	D 244	......	......	......	......	20–100	20–100
Furol Viscosity at 50°C, sec	T 59	D 244	20–100	100–400	50–500	50–500	......	......
Settlement, 5 days, % Difference	T 59	D 244	5−	5−	5−	5−	5−	5−
Sieve Test (Retained on No. 20), %	T 59	D 244	0.10−	0.10−	0.10−	0.10−	0.10−	0.10−
Aggregate Coating—Water Resistance Test	......	D 244						
Dry Aggregate (Job), % Coated			......	......	80+	80+	......	......
Wet Aggregate (Job), % Coated			......	......	60+	60+	......	......
Cement Mixing Test, %	T 59	D 244	......	......	......	......	2−	2−
Particle Charge Test	T 59	D 244	Positive	Positive	Positive	Positive	......	......
pH	T 200	E 70	......	......	......	......	6.7−	6.7−
Distillation:								
Residue, % by Weight	T 59	D 244	60+	65+	60+	65+	57+	57+
Oil Distillate, % by Volume of Emulsion	T 59	D 244	3−	3−	20−	12−	......	......
Tests on Residue								
Penetration, 25°C, 100g, 5 sec	T 49	D 5	100–250	100–250	100–250	100–250	100–200	40–90
Solubility in Carbon Tetrachloride, %	T 44	D 2042	97.0+	97.0+	97.0+	97.0+	97.0+	97.0+
Ductility, 25°C, cm.	T 51	D 113	40+	40+	40+	40+	40+	40+

TABLE 58.

Classification of Anionic Asphalt Emulsions According To B.S. 434-1960 and B.S. 2542-1960

Class	*Asphalt content (per cent)*	*Viscosity (degrees Engler at 20°C)*		*Application*
1. Labile or quick-breaking		Spraying-by bulk machine	6–13	
1 A	Nominal 62 (not less than 60)	Spraying-by hand operated machine	6–9	Surface dressing grouting
		Hand application-other than spraying	10–20	
1 B	Nominal 55 (not less than 53)		5–9	Grouting Surface dressing
1 C	Between 30 and 50 as agreed between supplier and purchaser		max. 5	Track coats Very thin film
2. Semi-stable				
2 A	Not less than 55		min. 15	Mixing*
2 B	Not less than 45		max. 15	"Retread" Process Mixing*
3. Stable	Not less than 50		max. 15	For all purposes**

* The choice of sub-class to be used depends on the nature and granding of the aggregate.

** Involving mixing with aggregates, including those containing large proportions on fines or chemically active materials such as cement, hydrated lime etc.

These groups can be made according to the setting value:

Rapid setting,
Medium setting, and
Slow setting groups

In accordance with the specification on the homogeneity of emulsions, 0.10% on 0.630 mm mesh sieve, and 0.25% on 0.160 mm mesh sieve are the maximum allowable. The adhesive power is also covered in the specification.

It is mentioned in the French specification that anionic emulsions may be used only for low category roads. The Asphalt Institute specifications for anionic asphalt emulsions can be seen in Table 56. Table 57 contains the specifications for cationic emulsified asphalts. The allowable large oil distillate quantity in the medium setting group refers to the solvent amounts used for its production. The classification of anionic emulsions commonly used in England as well as the data associated with the asphalt content and the uses are shown in a survey by Raudenbusch on Table 58[56]. Two types of cationic asphalt emulsions are still used in Germany at present:[57]

U 60 K to spray for surface treatment and similar purposes

M 65 K emulsions for mixing to produce macadam and similar materials

The rapid increase in the production and consumption of cationic emulsions rendered the development of suitable test and evaluation methods necessary[58,59].

3.2.4 Recommended Uses of Asphalts

After dealing with specifications for asphalt cement, liquid asphalt, and asphalt emulsions, the asphalt qualities to be used in the individual road construction techniques are summarized in Table 59 on the basis of a compilation prepared in the Asphalt Institute.

The liquid asphalt application is compiled in ASTM D 2399-65. This recommendation practice covers the selection of liquid asphalts of the slow, medium, and rapid curing types for various paving and related uses. Similarly, the special standards in each country refer to the suggested uses for a product, or the application is regulated by separate specifications covering the uses of the individual paving asphalts.

The problems arising in asphalt road construction and maintenance are dealt with in a great number of publications[60,61,62] all over the world. Besides, valuable information can be gathered from certain

TABLE
Recommended Uses

Type of Construction	References*	Paving Asphalts					Liquid — Rapid Curing (RC)			
Asphalt Concrete and Plant Mix, Hot Laid		40–50	60–70	85–100	120–150	200–300	70	250	800	3000
Highways	SS–1		×	×	×					
Airports	MS–11		×	×						
Parking Areas	MS–4		×	×						
Driveways	MS–4			×						
Curbs	SS–3			×[1]						
Industrial Floors	MS–4	×								
Blocks	MS–4	×								
PLANT MIX, COLD LAID										
Graded Aggregate	SS–1									
ROAD MIX										
Open-graded Aggregate	MS–14							×	×	
Dense-graded Aggregate	MS–14							×	×	
Clean Sand	MS–14							×	×	
Sandy Soil	MS–14						×	×	×	
PENETRATION MACADAM										
Large Voids	MS–13			×					×	×
Small Voids	MS–13				×			×		
SURFACE TREATMENTS										
Single, Multile and Aggregate Seal	MS–13				×	×		×	×	×
Sand Seal	MS–13							×	×	
Slurry Seal	MS–13									
Fog Seal	MS–13									
Prime Coat, open surfaces	MS–13							×		
Prime Coat, tight surfaces	MS–13						×			
Tack Coat	MS–13						×			
Dust Laying	MS–13									

59.
of Asphalt

Asphalts																			
Medium Curing (MC)					*Slow Curing (SC)*				*Emulsified*										
									Anionic					*Cationic*					
30	*70*	*250*	*800*	*3000*	*70*	*250*	*800*	*3000*	*RS–1*	*RS–2*	*MS–2*	*SS–1*	*SS–1h*	*RS–2K*	*RS–3K*	*SM–K*	*CM–K*	*SS–K*	*SS–Kh*
				×				×											
			×				×					×						×	×
			×	×							×						×		
		×	×			×	×				×	×				×		×	
		×	×										×					×	×
		×	×			×	×					×				×		×	×
										×					×				
									×					×					
		×	×	×					×	×		×		×	×			×	
		×	×						×			×	×	×				×	×
												×	×					×	×
												×[2]	×[2]					×[2]	×[2]
		×																	
×	×				×														
									×			×[2]	×[2]	×				×[2]	×[2]
×	×				×							×[2]						×[2]	

TABLE 59.

Type of Construction		*Paving Asphalts*					*Liquid*			
							Rapid Curing (RC)			
	*References**	*40–50*	*60–70*	*85–100*	*120–150*	*200–300*	*70*	*250*	*800*	*3000*
PATCHING MIX										
Immediate Use	SS–1									
Stock Pile	SS–1									
HYDRAULIC STRUCTURES										
Membrane Linings, Canals Reservoirs	MS–12	×[3]								
Hot Laid, Graded Aggregate Mix for Groins, Dam Facings, Canal Reservoir Linings	MS–12	×	×							
CRACK FILLING	MS–4						×			
MEMBRANE ENVELOPE	MS–1	×	×							
EXPANSION JOINTS	MS–4									
UNDERSEALING PCC	SS–6									
ROOFING	MS–4									
MISCELLANEOUS										

In northern areas where rate of curing is slower, a shift from MC to RC or from SC to MC may be desirable. For very warm climates, a shift to next heavier grade may be warranted.

1. In combination with powdered asphalt.
2. Diluted with water.

organizations such as AASHO, the HRB (Highway Research Board), and from various research institutes such as the British Road Research Laboratory and the French Laboratoires Central des Ponts et Chaussées, as well as from the Asphalt Institute Manual and Construction Series.

(Continued)

Asphalts																			
Medium Curing (MC)					*Slow Curing (SC)*				*Emulsified*										
									Anionic					*Cationic*					
30	70	250	800	3000	70	250	800	3000	*RS–1*	*RS–2*	*MS–2*	*SS–1*	*SS–1h*	*RS–2K*	*RS–3K*	*SM–K*	*CM–K*	*SS–K*	*SS–Kh*
		×					×												
		×	×			×	×									×	×		
									×			×[4]	×[4]	×				×[4]	×[4]

Blown asphalts, mineral-filled asphalt cements, and preformed joint com positions

Blown asphalts

Blown asphalts

Specially prepared asphalts for pipe coatings, battery boxes, automobile undersealing, electrical wire coating, insulation, tires, paints, asphalt tile, wall board, paper sizing, waterproofing, floor mats, ice cream sacks, adhesives, phonograph records, tree grafting compounds, grouting mixtures, etc.

3. Also 50-60 penetration blown asphalt and prefabricated panels.
4. Slurry mix.
* Publications of The Asphalt Institute where specifications or additional information may be found.

3.3 Asphalt for Roofing and Building

Asphalt, with its extraordinary combination of waterproofing, preservative, and cementing qualities, was an early discovery of ancient peoples. They used it extensively in the construction of their buildings, many of which, after thousands of years of exposure to the elements, are still existing, well preserved. Today asphalt is recognized more than ever as the outstanding roofing material and it is used extensively to cover all kinds of buildings, from homes to factories. Asphalt roofing materials are fully covered by specifications of the ASTM (62) and many government and private agencies. In addition, all types of asphalt roofing are sanctioned by Federal and other public agencies.

The large asphalt demand in building practice follows the amount used up in road construction. The bulk of these asphalts are blown asphalts with high softening points because of the service requirements. It should be mentioned that apart from roofs, building protective coating, flooring mastic etc., various asphalts, asphaltic solutions, emulsions, and other products are used in civil engineering and building practice.

Among the fields of application enumerated above, the asphalt demand for roofing is the most important. A separate group is represented by protective materials, where a great many products on asphalt basis are put on the market. A sharp separation of the qualities and uses of the various asphaltic materials used in building practice is not always possible, since asphalts standardized for road purposes are also used in building operations. These asphaltic solutions and compositions may be used successfully in other fields as well as in construction engineering.

In agreement with the above stated circumstances, the standard specifications of asphalts for roofing and building practice are different in various countries. The outstanding characteristics of Soviet building asphalt (GOST 9548-60) are summarised in Table 60. The asphalt groups shown here possess lower penetrations and higher softening points. These asphalt groups may be considered the continuation of road asphalts apart from the quality BN-VK.

The asphalts for roofing are standardized separately.

The function of roofing is to protect structures and their contents

TABLE 60.
Asphalts for Building Industry (GOST 6617-56)

Designation	*BN–IV*	*BN–V*	*BN–VK*
Penetration at 25°C, 0.1 mm	21–40	5–20	min. 20
Ductility at 25°C cm min.	3	1	—
Softening point, °C min.	70	90	90
Water soluble, % max.	0.3	0.3	0.3

from the elements and in some cases to add to the appearance of the building.

Asphalt roofs are of two basic types:

a) Built-up, that is, roofs built in place, consisting of plies of asphalt-saturated felt, asbestos felt, or fiberglass bound and protected by layers of asphalt and frequently topped with mineral cores

b) Prepared-roofs, that is roofs applied as finished, centrally manufactured roof products such as smooth roll roofing, granule-surfaced roll roofing, or granule-surfaced asphalt shingles

The quality of the asphalt used for the construction of built-up roofs should have sufficient resistance to flow so that it will not run off the roof to which it is applied. It should be waterproof and resistant to the effects of weather as long as possible. In order to insure good adhesion, it is desirable to use asphalt containing very little paraffin[(63)].

Various cementing materials are used in built-up roofing. On the basis of manufacturing methods, materials for hot application consisting of pure asphalt are distinguished from substances containing fillers. Besides, there are cementing materials for cold application which are made of asphalt and a suitable solvent. Primers made of asphalt and solvent are used in roofing work. Pigments and metal powders are applied when coloured materials are required.

The ASTM 312-64 standard specifications cover asphalt intended for use as hot-cement and mopping coat in the construction of built-up roof coverings for roofs surfaced in various manners, laid either over boards or concrete on various inclines. The asphalt qualities as per the standard are shown in Table 61. Mastic for roofing consisting of asphalt, filler, and containing antiseptics is specified in the Soviet standard GOST 2889-67. Asphalt with softening point of maximum 45, 50, 60, 70, and 85°C may be used for making the materials of the five groups. The filler quantities are from 15 to 25, and from 20 to 35%, respectively, depending on the fact whether fibre or combined fillers are used.

The heart of the entire built-up roofing operation is the kettle for heating the asphalt to the proper temperature for application. For efficient operation, the kettle must be able to bring the asphalt up to this temperature in a short period of time without overheating at the start, or at any time during the operation of the kettle. Overheating may seriously damage the asphalt and considerably reduce the life expectancy of the roof. Most of these overheating problems can be

TABLE 61.
Requirements for Asphalt for Constructing Built-up Roof Coverings ASTM D 312-64 (1958)

	I.		*II.*		*III.*		*IV.*	
Type of roofing	*min.*	*max.*	*min.*	*max.*	*min.*	*max.*	*min.*	*max.*
Softening point, °C	57	65	71	79	82	93	96	107
Penetration								
at 0°C, 200 g 60 sec	5	—	2	—	6	—	6	—
at 25°C, 100 g 5 sec	18	60	18	40	15	35	12	25
at 46°C, 50 g 5 sec	90	180	—	100	—	90	—	75
Ductility at 25°C, cm	10	—	3	—	3	—	1.5	—
Total bitumen soluble in carbon disulfide, %								
Mineral-stablized asphalt	65	—	65	—	65	—	65	—
Asphalts without mineral stabilize	99	—	99	—	99	—	99	—
Ash, per cent mineral stabilized asphalt	10	35	10	35	10	35	10	35

solved by the use of the recently developed thermostatically controlled kettles. The flow properties of the asphalt must be considered in the selection of asphalt for roofing. Associated with this, the difference between distilled and blown asphalts must be pointed out. If an asphalt is required in service that should not break at low temperatures and should not flow at high temperatures, blown asphalts will have to be used. However, the higher viscosity of the product must be considered when blown asphalt is used. It must therefore be heated to the desired temperature. Greatest care must be taken in production due to the rapid solidification of blown asphalt.

Cold process roofing is also used, where the hot process roofing is too hazardous or where simplification of application (without the necessity of heating the asphalt) and economical maintenance are primary considerations.

The materials in the application of a cold process are:

a) asphalt-saturated felt

b) adhesive or "cement" to bond the felt to the deck,

c) surfacing material

The asphalt used in such cements must have a sufficiently high softening point (typically an air blown asphalt of 90°C softening point) or stabilized with additional materials. Typical formulation of a cement: asphalt 50%, asbestos 10%, solvent 40%[64]. The top surfacing

TABLE 62.
Asphalt-base Emulsions for Use as Protective Coatings for Built-up Roofs ASTM D 1227-65

Type	*I.*		*II.*	
	min.	*max.*	*min.*	*max.*
Weight per gallon, lb	8.2	8.7	9.2	9.3
per liter, g	980	1040	1100	1140
Residue by evaporation %	45	55	40	60
Water, %	45	55	40	60
Ash, on basis of non volatile material, %	5	20	30	50
Non volatile organic base	asphalt		asphalt	
Inorganic reinforcement	asbestos		mineral fillers	

may be either a cutback or all emulsion. The ASTM D 1227-65 summarizes the asphalt-base emulsions for use as protective coatings for built-up roofs. These specifications cover asphalt-base emulsions capable of being spray or brush applied in relatively thick films as a protective coating for roof surfaces having inclines not less than 13 mm per 305 mm.

The requirements for emulsions and compositions are shown in Table 62. Type I includes asphalt-base emulsions prepared with mineral colloid emulsifying agents, Type II with chemical emulsifying agents.

Prepared roofing

a) Asphalt shingles. This type is composed of three basic materials: asphalt, felt and mineral granules. The felt is impregnated with an asphalt saturant. Both sides of the saturated felt are then covered with a harder, tougher coat of asphalt, which is then cut into individual shingles or strips.

b) Asphalt roll roofing-smooth surfaced. This type is made up of a single layer of roofing felt saturated with asphalt and coated on both sides with a harder asphalt.

c) Asphalt roll roofing-mineral surfaced. This type consists of smooth surfaced roll roofing in which mineral granules have been embedded on either one or both sides.

Roofing manufacture is an integrated, continuous line operation employing several unit operations. The most important of these unit operations are impregnating or saturating the felt, mixing stabilizer with coating asphalt, applying the stabilized coating, applying and embedding the granules, embossing a pattern, cooling, cutting and packaging. The saturation operation is an important and critical step. The saturation temperature and the asphalt viscosity connected with the former have to be set with care. This in turn renders necessary the thorough knowledge of asphalt viscosity-temperature relationship. Since correct saturation is influenced by the temperature, the saturation time, and other circumstances, all these factors must be considered to select optimum operation conditions.

The coating asphalt provides a heavy continuous roofing layer that is durable and resistant to the elements and it serves as binder for the granules. The function of the stabilizer is to add mass to the coating, serve as an extender and improve the coating durability.

The roofing sheet quality is covered by the Soviet standard GOST 2697-64. Quality requirements of roofing asphalts are summarized in GOST 9548-60. They are prepared from crude oil residue by blowing in accordance with the standard. The relating data are given in Table 63. The quality BNK-2 of the standard is used to saturate sheets.

No special building asphalts are standardized in Germany. For the above mentioned various purposes, road asphalts, the commonly used various blown asphalts and also high vacuum asphalts are utilized, if necessary.

Soft asphalt is used to impregnate the felt in the roofing manufacture, depending on the temperature load at the building plot; qualities B 200 and B 80 will generally meet the requirements. The impregnated material is coated by elastic blown asphalt in the course of roofing. The softening point of the blown asphalt applied amounts to 70-90°C and its penetration varies between 15 and 40 at 25°[65].

Asphalt insulating siding affords protection to the sidewalls of buildings. Asphalt siding is made in sheets or shingle form. Siding industry employs for saturation asphalts with high temperature susceptibility. These are usually of thermal cracked origin. The coatings are the

TABLE 63.
Roofing Asphalts (GOST 9548-60)

Designation	*BNK–2*	*BNK–5*
Penetration at 25°C, 0.1 mm	140	20
Softening point, °C min.	40	90
Loss on heating at 160°C, 5 hrs, max. %	1.0	1.0
Penetration decrease after loss on heating		
Water soluble, % max.	0.3	0.3

TABLE 64.
Requirements for Asphalt for Dampproofing and Waterproofing ASTM D 499-49 (1965)

	A		*B*		*C*	
Type	*min.*	*max.*	*min.*	*max.*	*min.*	*max.*
Softening point, °C	46	63	63	77	82	93
Penetration, 0.1 mm						
at 0°C, (200 g 60 sec)	5	—	10	—	10	—
at 25°C, (100 g 5 sec)	50	100	25	50	20	40
at 46°C, (50 g 5 sec)	100	—	—	115	—	100
Ductility at 25°C, em	30	—	15	—	2	—

blown asphalt type, similar to those employed in roofing asphalt coatings.

Asphalts for waterproofing and damp-proofing are quite similar to those used in construction built-up roofs, but ASTM has separate specifications for this purpose.

The ASTM specifications (D 449-49) cover three types of asphalts suitable for use as a mopping coat in damp-proofing, or as a plying or mopping cement in the construction of a membrane system of waterproofing. Some data of the three asphalt types are given in Table 64.

Type A-A soft, adhesive, "self-healing" asphalt, which flows easily under the mop and which is suitable for use below ground level under uniformly moderate temperature conditions both during the process of installation and during service. This type of asphalt is suitable for foundations, tunnels, subways, etc.

Type B-A somewhat less susceptible asphalt with good adhesive and "self-healing" properties for use above ground level where not exposed to temperatures exceeding 52°C. This type of asphalt is suitable for railroad bridges, culverts, retaining walls, tanks, dams, conduits, and spray decks.

Type C-An asphalt less susceptible to temperature than type B, with good adhesive properties for use above ground level where exposed on vertical surfaces in direct sunlight or at temperatures above 52°C.

The protective coatings for buildings are necessary for the protection against the penetration of moisture and for waterproofing without hydrostatic pressure, first of all used for the paint of masonry.

Asphalt proper is seldom used without further treatment as building protective coating. An exception is its utilization for cementing or for simple hot paint. The treatment aims at altering the deformation under the influence of heat. Formulations of asphalt with numerous other materials are now known to meet the special requirements and for the purpose of changing certain asphalt properties[66].

In principle the manufacture of products containing various solvents, and solvents as well as fillers, as used in building practice, is rather simple. A solution of the desired viscosity is prepared from the selected asphalt and solvent, then the filler is added to an extent enabling the setting of the desired product consistency. With one asphalt-solvent mix, usually three to four compositions of various consistencies may be obtained[67]. Should the production take place batchwise, the filler quantity fixes on the basis of previous laboratory and plant tests is placed in a vessel equipped with a suitable mixer. Before stopping production, a sample has to be taken to determine the consistency, since in the manufacture of large amounts the properties and the quantity as well as the distribution of the filler will influence the consistency of the product. In the case of continuous production, the material containing the solvent is pressed into a mixing device, where it is mixed with the measured quantity of the filler. The filler is best transported to the mixer by means of a screw conveyor. Various materials may serve as fillers; in general asbestos proved to be best. The manufactured products range from liquids easy to spray up to high viscosity grease like consistencies.

Full details are given by Hoiberg[68] on a plant for the production of asphalt base protective coating materials. The layout of the plant, the techniques used, and the type of machines are stated.

Hot and cold application building protection materials are distinguished depending on the preparation method. Beside products containing solvents, products made on an emulsion basis are also used. The various priming paints and the paints proper belong to this group[69]. Straight reduced asphalts are used as building protectives, whenever no great temperature loads are to be expected at the users'. If the temperature stress is great, blown asphalts will be used. High vacuum asphalt is employed to ensure materials free from tackiness at ambient temperatures. A special group is represented by asphalt filler materials for joints, pipe coating etc. within the frame of building protection.

The asphalt mastic of 12, 16, or 22% asphalt contents also belongs to the asphaltic building protective materials, as used for paving. This is applied for roofing, flooring, bridge insulationsland foundation insulation against water pressure. Some qualities are standardized in several countries.

British Standards have, in recent years, played an increasingly important part in providing specifications for the building industry and for its related professions. Part of these deals with the possibilities of asphalt mastic application. The summary given by the Mastic Asphalts Advisory Council contains directives as to asphalt mastic application, considering the standards[70].

The specifications of ASTM D 491-65 relate to the asphalt mastic for use in water proofing. These specifications cover materials for asphalt mastic suitable for use in waterproofing, consisting of asphalt cement, mineral filler, and mineral aggregate. The properties of the asphalt cement to be used are given by the standard.

The hydraulic application is a very important feature in the use of asphalt in building. Some of these uses are: linings for canals, reservoirs, ditches, storm drains and waste treatment systems, revetments for stream erosion control, beach and lake erosion control structures and dam facings. The hydraulic application of asphalt is throughly dealt with by Asbeck[71] and Rose[72–73]. According to Burnett[74], asphaltic materials play an important part in hydraulic construction made by the Bureau of Reclamation due to low relative cost, general availability, and numerous varieties. Mainly the types B 200 and B 45 are utilized in hydraulic engineering, especially as mixes with mineral materials. B 300 serves as a pure asphalt cementing agent applied in a hot state between two layers, or for the pretreatment of the aggregate. Blown asphalts are applied in hydraulic engineering only in special cases. Since they remain fully elastic also at low temperatures, they are used primarily to produce membranes. Mainly the type 85/40 is taken into consideration for this purpose. ASTM D 2521-66 T contains asphalt specifications for use in waterproof membrane construction for canal, ditch or pond lining. These specifications cover an oxidized asphalt. The requirements are compiled in Table 65.

An increasing amount of high softening point asphalt is used in building for flooring and other purposes (B 15, HVB 85/95). The various applications of asphalt in building practice are summarized in Tables by Georgi[75]. Köves and Zakar[76] report on the basis of tests on animals that some asphalts are suitable for flooring in cow-sheds. The determination of phenol quantities, which are harmful in the case of flooring of this type, was investigated by Zakar and Markó[77].

TABLE 65.
Requirements for Asphalt for Use in Waterproof Construction, Membrane Construction ASTM D 2521-66 T

Requirements	
Softening point, °C	79–83
Penetration of original Sample:	
at 0°C, 200 g 60 sec	min. 30
at 25°C, 100 g 5 sec	50–60
at 46°C, 50 g 5 sec	max. 120
Ductility at 25°C, cm	min. 3,5
Flash point (Cleveland open cup), °C	min. 218
Solubility in carbon tetrachloride, %	min. 97,0
Loss on heating, %	max. 1,0
Penetration at 25°C after loss on heating, of original, %	min. 60

3.4 Miscellaneous Uses of Asphalt

Beside the application of asphalt in paving and in building as well as roofing industry, it is used in a great number of industries and in many fields due to its favourable properties and low price. The amount of asphalt utilized for these special uses represents only a small part of the total asphalt consumption. Nevertheless the knowledge of the right selection and the use of the suitable quality is important enough.

Huge corrosion losses are prevented by the application of asphalt. Metal structures exposed to air, to humidity, or laid underground without protection will corrode. Several methods are known to insure corrosion resistance. One of the oldest processes is painting and varnishing, respectively. Asphaltic coatings may also be included into this group. The metal is separated from the medium causing corrosion by means of the asphalt coating. Special care must be taken to apply the protective coating in the necessary thickness and especially they should be non-porous. Asphalt has a medium position in corrosion protection as compared with plastics[78], and the low price here still represents a substantial advantage.

Three groups are distinguished in the application of asphalt for protection against corrosion:

1) General protection of metal structures by coatings
2) Protective coatings for pipe lines
3) Protection against corrosion in the cable and electric industries

The corrosion protection in construction industry is not dealt with here separately, since this belongs to the application of asphalt in building constructions.

Asphalt is used for the above purposes in a pure state as hot applied coatings, or "cut" by solvents, as well as in the form of emulsions as cold applied coatings. The bulk of the products utilized in this field contain fillers and additives, similar to the asphaltic materials used in the building industry. The asphalts utilized are blown qualities[79,80]. The British Standard Specification BS 4147-1967 for asphalt based coatings for ferrous products covers asphalt based materials, both unfilled and reinforced with powdered or fibrous fillers, for hot application which will provide effective coating for the protection of iron and steel.

Three main types are covered, with subdivisions for different conditions of use,—Type 1 unfilled compounds, grades *a*, *b*, *c*, and *d*—Type 2 and Type 3, compounds with or without mineral fillers. The characteristics of unfilled compounds are shown in Table 66. Grade *d* is suitable for use for the coating by hot dipping of articles which have not received a priming coat but which have been heated before immersion in the dipping bath. The asphalt *d* is required for charging the dipping bath and for its replenishment.

The filler contents of Type 2 and Type 3 vary between 20 and 60%. The British Standard also specifies the required characteristics of primers for asphalt coatings for hot application. The primer is applied as a thin film to metal in order to ensure maximum adhesion of the subsequent protective coatings. The asphalt based primers for cold application consist of a homogeneous solution of asphalt in white spirit. For hot application pure asphalt is used, as seen in Table 67.

TABLE 66.
Characteristics of Unfilled Compounds (British Standard 4147:1967)

	Type 1			
Grade	*a*	*b*	*c*	*d*
Softening point, °C	80–100	100–120	120–140	min. 40
Specific gravity, g/ml at 25°C, max		1,06		
Penetration at 25°C, 0.1 mm	18–45	10–25	5–20	max. 80
Solubility in CS, %		min. 99		
Loss on heating, % max.	0.5	0.5	0.5	1.0

TABLE 67.
Asphalt Based Primers for Hot Application (British Standard 4147:1967)

Characteristics		
Softening point, °C	80–100	100–120
penetration at 25°C, 0.1 mm	20–30	10–20
Solubility in CS_2, % min.	99	99
Loss on heating, % max.	0.2	0.2

These are asphaltic materials for cold solutions at the same time.

Each country possesses her own specifications for the insulation of pipe lines and for the quality of the asphalt required for it. Recent investigations and specifications deal with corrosion protection problems associated with the huge petroleum and gas pipe line constructions[81,82]. The referring asphalt groups comprising blown asphalts according to the Soviet standard GOST 9812-61 are shown in Table 68.

Several special tests are provided for in this standard. Besides, the paraffin wax content is shown in percentage for the first time. The Dutch "Communication No. 13" is the oldest and at the same time the most comprehensive specification issued in a European country[83] for the protection of underground pipes. Formerly also reduced asphalts were allowed for use to impregnate various coating materials, but blown asphalts are required by the present specifications in all these cases. In Germany[84] the following blown asphalts used in the country are applied in protective coatings and materials for pipe protection: 85/25, 85/40, and 115/15. Detailed specifications were issued

TABLE 68.
Asphalt for Gas-and Crude Oil Pipelines (GOST 9812-61)

Designation	*BNI–IV*	*BNI–IV–Z*	*BNI–V*
Penetration at 25°C, 0.1 mm	25–40	30–40	min. 20
Ductility at 25°C, max. cm	4	4	2
Softening point, min. °C	75	65 a. 75	90
Water soluble, max. %	0.2	0.2	0.2
Saturation in 24 hrs, %	0.2	0.2	0.2
Asphaltic acids, min. %	1.25	1.25	1.25
Paraffin wax, max.	—	4	—
Sulfur, max. %	—	2	—

TABLE 69.

Asphalt Enamel (Wrapped Systems, Specification M-1)
The Asphalt Institute Constr. Ser. No. 96

Grade	*A*	*B*
Softening point, °C	99–116	116–126
Penetration at 25°C max.	14	7
at 46°C 50 g 5 sec. min.	5	8
Flash point (Cleveland Open. Cup) °C, min.	232	232
Loss on heating, 5 hrs, %, max.	0.5	0.5
Ash, %	10–40	10–40

in the USA by the NACE (National Association of Corrosion Engineers) and by the Asphalt Institute(85,86).

As recommended by the Asphalt Institute (Construction Series No. 96), the asphalt protective coatings for pipe lines are divided into three major types:

a) Wrapped Systems
b) Mastic Systems
c) Coating for Interior Surfaces

Asphalt wrapped systems for pipe lines consist of a prime coat followed by either one or two applications of asphalt enamel in conjunction with one or more layers of reinforcing and protective wrappings. An outer wrap may sometimes be applied in place of or in addition to the inner wrap. The asphalt enamel shall be composed of asphalt combined with appropriate inert mineral fillers. It shall meet the requirements given in Table 69. Apart from this, the specifications of the Table cover the settlement, flow resistance and electrical resistance.

TABLE 70.

Asphalt Binder (Mastic Systems, Specification M-2)
The Asphalt Institute Constr. Ser. No. 96

Grade	*I*	*II*	*III*	*IV*
Operating temperature, °C	27	52	71	99
Softening point, °C	66–79	79–93	93–104	121–129
Penetration at 25°C 0.1 mm	21–25	15–17	7–11	5–8
Flash point (Cleveland Open Cup), °C, min.	232	232	232	232
Ductility at 25°C, min. cm	3.5	2.5	1.0	0

TABLE 71.
Asphalt Enamel (Coatings for Interior Surfaces, Specification M-3)
The Asphalt Institute, Constr. Ser. No. 96

Requirements	
Softening point, °C	116–126
Penetration at 25°C, max. (0.1 mm)	7
at 46°C, (50 g 5 sec.) min.	3
Ash, %	10–40
Settlement, max	2:1
Flow resistance	
Penetration at 29°C 100 hrs, inches, max.	0.01
at 46°C, 6 hrs, inches, max.	0.02
Electrical resistance, salt water immersion, 7 days, megohms/ft^2, min.	1000

Mastic systems for pipe lines consist of a prime coat followed by a coating of a dense, impervious, essentially voidless mixture of asphalt, mineral aggregate, and mineral filler, which may include asbestos fiber.

The asphalt binder for mastic systems shall be asphalt containing no mineral matter other than that naturally present in the asphalt. The individual asphalt binder grades are shown in Table 70.

Asphalt coatings for interior surfaces of pipe lines consist of a prime coat followed by a centrifugally cast layer of asphalt enamel. The asphalt enamel shall be composed of asphalt combined with appropriate inert mineral fillers. The most important requirements are compiled in Table 71. Apart from the Table, the specifications cover also the settlement, the flow resistance, and electrical resistance.

Among the specifications of producers and users, generally corresponding to the requirements of many different specifications[87], data on Shell specifications will be given[88]. Table 72 shows the Pipe Asphalt specifications of Shell. Eight qualities of the Carboplast enamel types are used in France[89], depending on circumstances.

Cold applied asphaltic coatings should not be considered only as a convenient way of applying a coating that would otherwise have to be melted, for they often can be formulated to do a job completely impractical or even impossible with a hot-melt coating. The two methods for modifying asphalt so that it may be applied cold are similar to those used by the paint industry to apply various resinous coatings. They may be dissolved in a solvent or made into an aqueous emulsion.

TABLE 72.
Shell Pipe Asphalts Specifications

Types		"*CC*"	"*BPM*"	"*H*"
Softening point R B, °C	ASTM D36	90 minimum	122 to 130	113 to 118
Penetration at 25°C 100g/5 sec, 0.1 mm	ASTM D5	—	10 to 20	4 to 7
Penetration at 66°C 50 g/5 sec, 0.1 mm		—	—	30 maximum
Flow at 70°C, 20 hr, mm 5 mm	Communication	6 maximum	—	—
Flow at 100°C, 20 hr, mm layer at 45 angle	13, Test 3.4	—	30 maximum	—
Flow resistance, in 2 lb/in^2 at 46°C, 6 hr	Asphalt Insti- tute, Method 12	—	—	0.015 maximum
Indentation test 2.5 kg/cm^2 mm, at 24 hr	Communication Test 3.5	13, 17 maximum	10 maximum	—
Shatter test at 0°C, g	Communication	13, 66.8 minimum	—	38.5 minimum
Shatter test at 25°C, g	Test 3.6 d	—	225 minimum	—

A long line of useful products is produced by both methods, with each type having its own distinct characteristics and uses[90].

In the cable and electric industries, asphalt is used as an anticorrosion protective coating on the external metal jacket of the cable as well as filling compound, sealing cables and other armatures, further in

TABLE 73.
Electric Industry Asphalt (Hung. Standard 3266-57)

Designation	K_1	K_2	K_3
Penetration at 25°C, 0.1 mm	50–100	20–60	20–40
Softening point, °C	49–54	59–65	80–100
Brittleness and adhesivity	should meet the requirements of the lead stripe test*		
	at +5°C	at +10°C	at +25°C
Phenolic reaction		negative	
Ductility at 25°C, cm	100	15	2
Paraffin wax, max. %	2.5	2.5	2.5

* 1 mm thick asphalt layer applied to a 170 mm long, 14 mm wide, and 0M9 mm thick lead stripe has to be wound on a glass or metal cylinder of 10 mm diameter.

TABLE 74.
Asphalt for Battery Mastic GOST 8771-58

Requirements	
Penetration at 25°C 0.1 mm	8–11
Ductility at 25–C, cm min.	1.5
Softening point, °C	102–110
Water soluble, % max.	0.3
Water soluble acid and base	none
Ash, max. %	0.4

various asphaltic insulation varnishes. The Hungarian standard for electric industry asphalt (MSz 3266-57) includes three asphalt qualities, the outstanding specifications of which are shown in Table 73.

Quality K_1 is obtained by distillation, while K_2 and K_3 are made by blowing. A special specification is the determination of brittleness and that of adhesivity, for which the lead stripe test is used. In Germany, asphalt B 300 is used for impregnating cable wrapping tape, and blown asphalt qualities 75/30 and 85/25 are taken for cable filling compound.

The specifications of the Soviet standard GOST 8771-58 for filling compound asphalt are given in Table 74. Formerly nearly all asphalt varnishes were prepared from native asphalts. The designation used up to now originates from this. Increasing amounts of asphalts obtained by reduction from crude oils are now used in lacquer and varnish production. Asphalts are commonly somewhat softer, therefore they are often treated mixed with native asphalts[91].

Straight run and blown asphalts of high softening points are used in the varnish industry. Thus the asphalts utilized in most cases are high vacuum asphalts with softening points 130/140 or higher and blown asphalts of 135/10 and 160/5 qualities,[92]. Table 75 illustrates the outstanding data referring to the asphalt qualities in the Soviet varnish

TABLE 75.
Asphalts in the Varnish Industry (GOST 3508-55)

Designation	*B*	*V*	*G*
Softening point, °C	100–110	110–125	125–135
Penetration at 25°C 0.1 mm	11	8	5
Benzene insoluble, % max.	0.2	0.2	0.2
Ash, % max.	0.3	0.3	0.3
Acid value mgKOH/g asphalt	2.0	2.0	2.0

industry[93]. One of the most important specifications is the special claim of its being miscible with linseed oil.

The insulating and water resistant properties of asphalt are very valuable also in the paper industry[94] The papers are usually saturated with asphalt and may also be coated or may consist of two or more thicknesses of dry paper cemented together with asphalt.

One type belonging to insulation and waterproof papers belongs to the group of asphalt-treated papers. They can be divided into the following groups:

1) Asphalt saturated papers
2) Papers coated on one side
3) Laminated papers

Asphalt cements with penetrations at 25°C of 200, 80, and 65, are used for impregnation.

To produce a paper coated on one side, non-adhesive hard asphalt is suitable. High vacuum asphalts of 85/95 and 95/105 softening points proved to be satisfactory for this purpose. Laminated paper is commonly made of blown asphalt 85/25 and 85/40. Other qualities are used for different purposes in the paper industry beside those enumerated, e.g. asphalt emulsions.

Asphalt is known as "mineral rubber" in the rubber industry. Formerly the marketed products were native asphalt types with high softening points, but a 160/5 blown asphalt is the quality commonly used. Blown asphalts of 85/40, 100/15, 115/15, and 135/10 are also used to produce certain rubber products[95]. The softening point of the standardized quality was established from 120 to 140°, the penetration between 10 and 30 as a result of investigations on Hungarian asphalt from Nagylengyel for possible use in the rubber industry[96]. Ten to fifteen percent of rubber can be substituted on the basis of tests

TABLE 76.
Asphalts for Rubber Industry (Rubraks) GOST 781-51

Designation	*A*	*B*
Softening point, °C	125–135	135–150
Ash , max. %	0.8	0.8
Loss on heating during 2 hrs at 150°C, %	0.1	0.1
Water soluble acid	0	0
Water soluble base, max.	traces	traces

associated with the carbon black content of the rubber without involving the deterioration of the quality. The specifications for the blown asphalt for the rubber industry (Rubraks) standardised in the Soviet Union are compiled in Table 76.

In the pretreatment of raw rubber and synthetic rubber, a soft asphaltic product may be used as a rubber substitute, which may be considered a rubber substance proper during further treatment of the mix. The liquid asphalt of viscosity 20°E at 100°C is straight run from Nagylengyel crude oil; this proved to be satisfactory for such purposes.

Asphalt is a good binder for briquetting[97]. Large quantities of asphalt are used here instead of coal tar pitch in certain countries[98,99]. Asphalt quality depends on briquetting techniques[100].

Summarizing the relationships between asphalt characteristics and briquette, Lang[101] states that the most important property of a briquetting binder is its penetration over the range of temperatures to which the briquettes may be exposed during handling, transportation, and storage. Briquettes made with a low penetration asphalt as binder tend to be brittle and will break and/or abrade when handled, while briquettes made with a high penetration asphalt may soften and either deform or stick. No one penetration of the asphalt is ideal, and this should also be varied depending on whether it is to be used during the summer or during the winter.

According to Charbonnier and Visman[102], the optimum specifications for an asphalt binder used in the molten state for briquetting

TABLE 77.

Optimum Specifications for an Asphalt Binder Used in Briquetting in the Molten State by Charbonnier and Visman

Requirements	
Origin	cracking
Softening point, °C	65–70
Penetration	
max. at 46°C	35
min. at 66°C	120
optimum at 25°C	5–20
Susceptibility:	
ratio $\frac{\text{penetration at 66°C}}{\text{penetration at 46°C}}$ min.	4.5
Moisture, max.	0.5
Insoluble in benzine max. %	1
Conradson index, min. %	35

TABLE 78.
Various Asphalts of Briquetting Industry

Origin of crude	*Production method*	*Asphalt*	
		Softening point, °C	*Penetration at 25°C*
Venezuela	Distillation	72	10
	Cracking	85	2
	Propane residue	90	4
	Cracking	95	1
Hungary	Distillation	76	15
(Nagylengyel)	Distillation	92	8
	Propane residue	82	4
Austria (Matzen)	Blowing	75	14

typical bituminous coals are shown in Table 77. If the asphalt is added to the mix in a ground and cold state, the optimum softening point of the asphalt varies between 65-100°C, according to the nature and method of processing the crudes, the briquetting techniques, the nature of the coals, and the outside temperature.

Characteristic data of briquetting industry asphalts of various origin are given in Table 78[103]. The specifications NIMSz 138-67 covering the asphalt binders of the two different type briquetting plants in Hungary are shown in Table 79.

The quality BB-75 is used in plants applying the asphalt in a molten state. Quality BB-85 is commonly used in the crushed form, but also the quality BB-75 may be ground in the cold season. If it is too warm, the quality BB-95 will be applied. According to the investigations of Zakar[104], the propane asphalt obtained from Romashkino crude can be utilised in briquette manufacture. Characteristic data of the

TABLE 79.
Asphalts for Briquetting (NIMSz 138-67)

Designation	*BB 75*	*BB 85*	*BB 95*
Softening point, °C	71–80	80–90	91–100
Penetration at 25°C, 0.1 mm	12–20	max. 14	max. 10
Ductility at 25°C, cm	min. 3	max. 5	max. 3
Ash, max. %	0.5	0.5	0.5
Water content, %	0.2*	0.2*	0.2*
Flash point, Marcusson, min. %	260	280	280

* In the case of asphalt made according to the so-called "floating process" water content is not standarized.

TABLE 80.
Propane Asphalt From Romashkino

Designation	*A*	*B*
Softening point, °C	62	74
Penetration at 25°C, 0.1 mm	12	4
Ductility at 25°C, cm	100	1
Fraass breaking point, °C	+9	+13
Conradson carbon residue, %	27.4	29.5
Paraffin wax, %	1.6	1.5

utilised propane asphalt are shown in Table 80. Generally the B quality is used. The A quality could also be ground in the cold season. In summer it may become necessary to resort to blown asphalt from propane asphalt due to the heat in the plant operating on crushed asphalt. A blown asphalt from cracked residues proved to meet the requirements best in France, on the basis of investigations which had been carried out for years. They also succeded in obtaining large quantities of coking residue with low penetrations. The characteristic features of briquetting asphalt marketed by Shell in France[105] are given in Table 81. A summary of the data of briquetting asphalts commonly used in the USA and Canada are published by Krchma[106].

Beside the above mentioned fields of outstanding importance, there are still several other application possibilities for asphalts of various properties. Thus asphalt is also applied in fuse manufacture. The quality of the utilized asphalt depends on the manufacturing technique. Asphalt of softening points between 63 and 72°C and that of softening points between 80 and 100°C are specified for internal coatings and for external coatings, in the Hungarian standard in question (MSz 8546). Blown asphalt is used for special purposes. Low penetration

TABLE 81.
French Asphalt for Briquetting

Designation	*Mexphalte* 78/3	*Mexphalte* 95/1
Softening point, °C		
Ring and ball method	77–81	94–98
Kraemer-Sarnow method	69–73	80–84
Penetration, 0.1 mm		
at 25°C max.	5	1
at 40°C max.	17	10
Conradson carbon residue, min. %	34	40

TABLE 82.
Hot Neck Grease of Asphalt Base (DIN 51824)

Designation	*90*	*1000*	*135**
Softening point, %C	85–95	95–105	130–140
Specific gravity at 25°C, g/ml	1.03–1.10	1.03–1.10	1.03–1.07
Flash point, min. °C	325	325	275
Penetration at 25°C, 0.1 mm	4–12	3–8	min. 12

* Blown asphalt base

asphalt of high softening point is utilized to make phonograph records, contrary to the previous type.

The characteristic data of hot neck grease on asphalt base are given in Table 82, which serve the purpose of lubricating roller pins and their sliding bearings as well as that of hot rolling stand (DIN 51824).

Asphalt may also be used as a binder with coking of poorly baking coal types; the asphalt coking residue serves as anode paste in aluminum production. Further applications are the production of printing ink, skiing wax, and core binding agent. Asphalt is also used in agriculture, and for stabilizing slopes and flat areas adjacent to highways(107). A thin asphalt layer is applied to decrease losses in soil moisture. The protection during 5-6 weeks of an asphalt film formed in the covering of the beds (the seed) with an asphalt layer ensures a satisfactory effect at the time of germination and breaking through the soil. The prevention of excessive moisture losses may become very important, if dried soil is intended to become fertile again. Asphalt is also used to collect rain-water. A dried surface is covered by an asphalt coating from which the rain-water flows off and can be conducted to the territory in question. Products developed for the warm, semi-arid areas of south-western USA have been shown to have utility in cooler, more humid climates(108).

There are two accepted methods for using asphalt in the mulching processes. One is to apply the asphalt alone, in a spray film on the seeded area. The other method is to employ the liquid asphalt as a tie-down for straw or hay. Established procedures have proved that asphalt can be used successfully in both types of treatment. The asphalt commonly used in film spray is an asphalt emulsion. For the tie-down method any type of liquid asphalt and asphalt emulsion is used.

Recent reports deal with the stabilization of sandy downs and the fixation of slopes by petroleum products. This stabilization has to be effective as long as vegetation growth becomes strong enough to protect the downs in a lasting manner[109,110,111]. Asphalt proved to be a valuable and satisfactory material in radiation protection[112]. It is suitable also for coating radioactive sludges. According to the report of the Commissariat à l'Energie Atomique, one stage coating process gives the best results. The procedure consists of pouring hot asphalt into the sludge with a surface-active agent. The most satisfactory assphalt are Mexphalte 40/50 for sludges of medium activity and 80/100 for sludges of high activity.[113].

For the sake of completeness, the application of asphalt as a fuel should also be mentioned. The applicability is limited in this field due to the high viscosity of asphalt[114]. Asphalt may be utilized, however, without greater changes in the equipments for heating units using coal dust. In such cases pitch asphalt which can be sprayed readily will prove satisfactory[115].

References

1. Bitumen. *27*. 212 (1965). Bitumen, Teere, Asphalte, Peche, verwandte Stoffe *16*, 193 (1965).
2. Goppel, J. M.: Knoterus, J.: Proc. fourth World Petr. Cong. Sec. III/G, Rom (1955).
3. Bitumen *30*. 26 (US Bureau of Mines).
4. Lang. M.: Bitume-Actualités, Juin, 1966.
5. Jarman, A. W.: The Mexphalte Handbook, Shell-Mex and B. P. London (1963).
6. The Asphalt Handbook, The Asphalt Institute Maryland (1968).
7. Bitumen Taschenbuch. ARBIT Hamburg (1964). Bauverlag.
8. Reznák, L.: Mélyépitéstudományi Szemle. *15*, 545 (1965).
9. Nowicki, L.: Drogownictwo *21* 198 (1966).
10. Rincheval Materials de Stockage et Chauffage de Liants Hydrocarbones, Etablissements Rincheval, Soisy-s/Montmorency-Prospectus
11. Behälter mit eingebauter Ölumlaufheizung Westhydraulik-Becker Kg. Bonn-Lengsdorf.–Prospectus
12. Thermalöl-Generatoren, Maschinenfabrik Theodor Ohl KG Limburg-Lahn.–Prospectus
13. Stever, F.: Bitumen. *27*, 225 (1965).
14. Bitumen, Teere, Asphalte, Peche, verwandte Stoffe *16*, 310 (1965).

15. Monasturskij, O. V.: Avtomatizacija rozogreva bituma i mastic v stroizdat, Stroizdat Moscow (1964).
16. Bitumen, Teere, Asphalte, Peche, verwandte Stoffe *16*, 470 (1965).
17. Horn, R.: Bitumen, Teere, Asphalte, Peche, verwandte Stoffe *16*, 503 (1965).
18. Nüssel, H.: Bitumen. Mainz (1958). Hüthig und Dreyer.
19. Ewers, N.: Wissenschaftliche Zeitschrift der Hochschule für Bauwesen Leipzig. *2*, 5 (1962).
20. Nitsch, W. H.: Bitumen, Teere, Asphalte, Peche, verwandte Stoffe *17*, 12 (1966) *11*. 445 (1960).
21. Proceedings of the Association of Asphalt Paving Technologist 1956-1968 Annual issue.
22. Bitumen in civil engineering, RILEM International Symposium, Dresden (1968).
23. Stinsky, F.: Strasse und Verkehr *54*, 99 (1968).
24. Stinsky, F.: Strasse und Verkehr *54*, 583 (1968).
25. Zakar, P.; Sixth World Petr. Congr, Proceedings Sec. VI. 158 Hamburg (1963).
26. MSz 3276-68.
27. Zakar, P.: Mélyépitéstudományi Szemle *18*, 363 (1968).
28. Mihailov, V. V. RILEM International Symposium, Dresden (1968).
29. Kolbanovskaya, A. S.: RILEM International Symposium, Dresden (1968).
30. Zakar, P.: Országos Mélyépitóipari Konferencia elöadásai Budapest (1961).
31. Gagle, D. W.: Oil Gas J. No. *38*, 121 (1960).
32. Smith, V.: Petroleum Refiner. *39*, 211 (1960).
33. Traxler, R. N.: Asphalt. New York (1961). Reinhold.
34. Ewers, N.: Wissenschaftliche Zeitschrift der Hochschule für Bauwesen Leipzig, 1 (1961).
35. Krenkler, K.: Bitumen, Teere, Asphalte, Peche, verwandte Stoffe *15*, 447, 499, 551 (1964).
36. Krom, C. J.; Dormon, G. M.: VI. Wld. Petroleum Congr. Hamburg Sec. VI. Paper 14 (1963).
37. Krom, C. I., Dormon, G. M.: Proceedings Seventh World Petroleum Congr. Vol. VIII. 463, Elsevier (1967).
38. Ivanov, N.: Proceedings Seventh World Petroleum Congress Vol. VIII. Elsevier (1967).

39. Welborn, I. Y.: Proceedings Seventh World Petroleum Congr. Mol. VIII. 451 Elsevier (1967).
40. Izatt, I. O.: Proceedings Seventh World Petroleum Congress Vol. VIII. 275 Elsevier (1967).
41. Monismith, C. L.: Proceedings Seventh World Petroleum Congress Vol. VIII. 425 Elsevier, (1967).
42. Zakar, P., Simon, M.: Bitumen, Teere, Asphalte, Peche, verwandte Stoffe, *15*, 562 (1964).
43. Güsfeldt, K. H.: Bitumen, Teere, Asphalte, Peche, verwandte Stoffe *17*, 5 (1966).
44. Zakar, P.: Strasse und Tiefbau *16*, 601 (1962).
45. Specificationes for Asphalt Cements and liquid Asphalts, The Asphalte Institute, Maryland (1968).
46. Fascicule n°24, Cahier des Prescriptions Communes des Marches de Travaux Publics, Paris (1966).
47. MSz-3268-65
48. Simon, M., Zakar, P.: Bitumen, Teere, Asphalte, Peche, verwandte Stoffe *15*, 359 (1964).
49. Lee, A. R., Nicholas, J. H.: Journal I. P. 43 (1957).
50. Matthew, D. H.: Bitumen, Teere, Asphalte, Peche. *11*, 230 (1960).
51. Kleinert, H.: Wissenschaftliche Zeitschrift der Hochschule für Bauwesen Leipzig. *1*, 71. (1961).
52. Dinoram its uses in the roadmaking industry, Auby-Prochinor-Prospectus.
53. Adhesion in Road Making with Duomeen T, Armour Hess Chemicals LTD–Prospectus.
54. Bewährte und neuentwickelte Bitumenzusätze, TH. Goldschmidt A.G. Essen–Prospectus.
55. Duriez M.; Revue Generale des Routes N° 437, 438 (1968).
56. Raudenbusch, H.: Bitumen Teere, Asphalte, Peche, verwandte Stoffe *17*, 43 (1966).
57. Strassenbau mit Bitumenemulsionen und Kaltbitumen, "Strassenbau von A-Z" Schmidt Verl. (1965).
58. Ariano, R.: Strade. *42*, 99 (1962).
59. Bierhalter, W., Mauch, K.: Bitumen, Teere, Asphalte, Peche, verwandte Stoffe *16*, 483, 533 (1965).
60. Wallace, H. A., Martin, J. R.: Asphalt pavement engineering, New York (1967). McGraw-Hill.

61. Jeuffroy, G.: Conception et Construction des Chausees, Eyrolles, Paris (1957).
62. 1967 Book of ASTM Standars, Part II. ASTM Philadelphia (1967).
63. Thurston R. R. by Hoiberg A.I. Bituminous Materials Vol. II. New York (1965). Interscience.
64. Berry, G. W. by Hoiberg A. I.: Bituminous Materials Vol. II New York (1965). Interscience.
65. Bitumen, Arbit-Schriftenreihe 1. Hamburg (1961).
66. Walther, H.: Bituminöse Stoffe im Bauwesen. Heidelberg (1962). Strassenbau, Chemie und Technik.
67. Traxler, R.: Asphalt, New York (1961). Reinhold.
68. Shearon W., H., Hoiberg, A. I.: Ind. and Eng. Chem. *41* 2672 (1949).
69. Heinrich, Fr. J.: Bituminöse Anstrichstoffe. Garmisch-Partenkirchen (1950). Moser.
70. Application of Mastic Asphalte, Mastic Asphalte Advisory Council, London (1966).
71. Asbeck Van, W. F.: Bitumen in hydraulic engineering. Shell. London (1959).
72. Rose, D.: Bitumen im Wasserbau, BP Benzin und Petroleum AG. Hamburg (1963).
73. Rose, D.: Proceedings Seventh World Petroleum Congress Vol. VIII. 343, (1967). Elsevier.
74. Burnett, G. E.: Proceedings Seventh World Petroleum Congress Vol. VIII. 295, (1967). Elsevier.
75. Georgy, M.: Die Baustoffe Bitumen und Teer. Köln-Braunsfeld (1963). Rudolph Müller.
76. Köves, J., Zakar, P.: Bitumen, Teere Asphalte, Peche, verwandte Stoffe *10*, 393 (1959).
77. Zakar, P. Markó, B.: Bitumen, Teere, Asphalte, Peche, verwandte Stoffe *16*, 303 (1965).
78. Seymour, R. B.: Mod, Plastics, *38*. 29 (1961).
79. Gundermann, E.: Bitumen, Teere, Asphalte, Peche, verwandte Stoffe *17*, 94 (1966).
80. Temme, T.: Bitumen, Teere, Asphalte, Peche *16*, 551 (1965).
81. Micu, J.: Petrol si Gaze. *10*, 503 (1959).
82. Benes, V.: Strojirenstvi. *10*, 789 (1960).
83. Mitteilung 13(Ausg. 1962) Metallinstitut T.N.O. Delft. Bitumen *27*, 22 (1965).

84. Bitumen und Asphalt Taschenbuch. ARBIT-Bauvelag. Wiesbaden (1957).
85. Nace Technical Committee Reports Publ. 57-11, 57-14, Corrosion 13, Apr. 75, May 77 (1957).
86. Asphalt protective coatings for pipe lines. The Asphalt Institute, Cónstr. Series 96.
87. Alexander, S. H., Tarver, G. W.: Ind. Eng. Chem. *58*, April, 37 (1966).
88. Pipeline Protection with Shell, Shell Construction Service–Prospectus
89. Protection des Canalisations Souterraines, Bitumes Speciaux, Paris–Prospectus
90. Dickson, W.I.: Ind. Eng. Chem. *58* April, 28 (1966).
91. Kisselew, W. S., Abzschkina, A. F.: Herstellung von Lacken, Ölfirnissen und Farben. Leipzig (1957). Fachbuchverlag.
92. Gundermann, E.: Chem. Techn. *11*, 441 (1959).
93. GOST 3508-55.
94. Ohl, F.: Impränieren von Papier und Pappe, Wiesbaden (1959). Dr Sändig Verl.
95. Kirchhof, F.: Bitumen, Teere, Asphalte, Peche. *1*, 64 (1950).
96. Kirchhof, F.: Bitumen, Teere, Asphalte, Peche. *1*, 64 (1950).
96. Bruckner, Z., Zakar, P.: a. o.: Magyar Kémikusok Lapja, *11*, 282 (1956).
97. Zakar, P., Tóth, J.: Brennstoff-Chemie. *38*, 373 (1958).
98. Chaix, R.: Revue de l'Industrie minérale. *42*, 103 (1960).
99. Zakar, P., Mozes, Gy.: Bitumen, Teere, Asphalte, Peche. *9*. 275 (1958).
100. Charbonnier, I., Lusinchi, I.: Proceedings International Briquetting Association, 47. NRRI University of Wyoming 1955.
101. Lang, W. A.: Proceedings International Briquetting Association, 2 NRRI University of Wyoming (1955).
102. Charbonnier, R. F., Wisman, I.: Proceedings International Briquetting Association, 12 NRRI University of Wyoming (1959).
103. Zakar, P., Aszmann G., Tóth, J.: Proceedings International Briqueting Association, 93 NRRI University of Wyoming (1959).
104. Zakar, P.: Bitumen, Teere, Asphalte, Peche. *18*, 438 (1967).
105. Societé des Petroles Shell Berre, Paris (Bitumen type agglomeration).

106. Krchma, L. C., by Hoiberg, A. J.: Bituminous Materials Vol. H. New York (1965), 583. Interscience.
107. Petroleum *22*, 338 (1959).
108. Neblett, R. F.: Proceedings Seventh World Petroleum Congr. Vol VIII. 251 Elsevier (1967).
109. Les, T. L., Richard, G. P.: Sixth World Petr. Congr. Proceedings Hamburg (1963) Sec. VI. Paper 10.
110. Bielfeldt, H. Taeubner, K.: Bitumen. *28*. 116 (1966).
111. Vogel, G.: Bitumen *28*, 116 (1966).
112. Ziehmann, G.: Bitumen *24*, 260 (1962).
113. Journal of the Institute of Petroleum *51*, 673 ref. (1965).
114. Lachaux, P.: Bull. Assoc. franc. Techniciens Pétrole. 31, Jan. 3. (1955).
115. Beuther, H., Goldthwait, R.: Petroleum Refiner. *41*. 12, Nr. 96 (1962).

Appendix

A—1
Authoritative Methods of Test as Commonly Specified

Art.		*AASHO*	*ASTM*
	Asphalt Cement		
3.02	Penetration	T49	D5
3.03	Viscosity (See also Saybolt Furol Test, at High Temperatures, or ASTM Method of Test E102)	T201	D2170
3.04	Flash Point (See also Pensky-Martens Flash Point Test, AASHO Method of Test T73 and ASTM Method of Test D93)	T48	D92
3.05	Thin Film Oven Test	T179	D1754
3.06	Ductility	T51	D113
3.07	Solubility	T44	D4
3.08	Specific Gravity	T43	D70
3.09	Softening Point	T53	D36
	Rapid-Curing and Medium-Curing Asphalt		
3.10	Viscosity (See also Saybolt Furol Test, AASHO Method of Test T72 or ASTM Method of Test D88	T201	D2170
3.11	Flash Point	T79	D1310
3.12	Distillation	T78	D402
3.13	Tests on Residue (See Asphalt Cements, Articles 3.02 thru 3.09)		
3.14	Water in Asphalt	T55	D95
3.15	Specific Gravity	T43	D70
	Slow-Curing Asphalt		
3.16	Viscosity (See RC and MC Asphalts)	T201	D2170
3.17	Flash Point (See Asphalt Cements)	T48	D92
3.18	Distillation	T78	D402
3.19	Float Test	T50	D139
3.20	Asphalt Residue of 100 Penetration	T56	D243
3.21	Ductility	T51	D113
3.22	Solubility (See Asphalt Cements and RC and MC Asphalts)	T44	D4
3.23	Water in Asphalt	T55	D95
3.24	Specific Gravity	T43	D70
	Emulsified Asphalt		
3.25	Viscosity	T59	D244
3.26	Residue from Distillation	T59	D244
3.27	Settlement	T59	D244

A—1 (Continued)

Art.		*AASHO*	*ASTM*
3.28	Demulsibility	T59	D244
3.29	Sieve Test	T59	D244
3.30	Cement Mixing	T59	D244
3.31	Tests on Residue (See Asphalt Cements, Articles 3.02 thru 3.09)		
3.32	Coating Test	—	D244–61T
3.33	Particle Charge Test	T59A	D244
3.34	pH Test	T200	E70
3.35	Oil Distillate	T59	D244
3.36	Specific Gravity	T43	D70

A—2
Temperature Conversion Table

°*C*	°*F*	°*C*	°*F*
−30	−22.0	+170	+338.0
−25	−13.0	+175	+347.0
−20	− 4.0	+180	+356.0
−15	+ 5.0	+185	+365.0
−10	+14.0	+190	+374.0
− 5	+23.0	+195	+383.0
0	+32.0	+200	+392.0
+ 5	+41.0	+205	+401.0
+10	+50.0	+210	+410.0
+15	+59.0	+215	+419.0
+20	+68.0	+220	+428.0
+25	+77.0	+225	+437.0
+30	+86.0	+230	+446.0
+35	+95.0	+235	+455.0
+40	+104.0	+240	+464.0
+45	+113.0	+245	+473.0
+50	+122.0	+250	+482.0
+55	+131.0	+255	+491.0
+60	+140.0	+260	+500.0
+65	+149.0	+265	+509.0
+70	+158.0	+270	+518.0
+75	+167.0	+275	+527.0
+80	+176.0	+280	+536.0
+85	+185.0	+285	+545.0
+90	+194.0	+290	+554.0
+95	+203.0	+295	+563.0
+100	+212.0	+300	+572.0
+105	+221.0	+305	+581.0
+110	+230.0	+310	+590.0
+115	+239.0	+315	+599.0
+120	+248.0	+320	+608.0

A—2 (Continued)

°C	°F	°C	°F
+125	+257.0	+325	+617.0
+130	+266.0	+330	+626.0
+135	+275.0	+335	+635.0
+140	+284.0	+340	+644.0
+145	+293.0	+345	+653.0
+150	+302.0	+350	+662.0
+155	+311.0	+355	+671.0
+160	+320.0	+360	+680.0
+165	+329.0	+365	+689.0

A—3
Table of Conversions

Units	*Metric Equivalents*	*Metric Units*	*Equivalents*
LINEAR MEASURE			
1 inch	2.54 cm = 0.0254 m	1 cm	0.394 in.
		1 m	39.4 in.
1 foot	30.5 cm = 0.305 m	1 m	3.3 ft.
1 yard	91.4 cm = 0.914 m	1 m	1.1 yd.
1 mile	1609 m	1 km	1094 yds.
SQUARE MEASURE			
1 sq.inch	6.45 sq.cm.	1 sq.cm.	0.155 sq.in.
1 sq.foot	0.093 sq.m.	1 sq.m.	10.76 sq.ft.
1 sq.yard	0.84 sq.m.	1 sq.m.	1.2 sq.yd.
1 acre	0.405 hectare	1 hectare	2.47 acres
CUBIC MEASURE			
1 cu.foot	0.028 cu.m.	1 cu.m.	35.3 cu.ft.
1 cu.yard	0.765 cu.m.	1 cu.m.	1.31 cu.yd.
CAPACITY MEASURE			
1 imperial gallon	4.54 litres	1 litre	0.22 imperial gallons
1 U.S.A. gallon	3.78 litres	1 litre	0.26 U.S.A. gallons
WEIGHT			
1 ounce	28.35 grammes = 0.02835 kg.	1 gr.	0.035 ounces
1 lb.	0.45 kg.	1 kg.	2.20 lb.
1 ton	1.016 tons	1 ton	0.984 tons

Index